W9-BZL-619

OXFORD LOGIC GUIDES

General Editors

DOV GABBAY
ANGUS MACINTYRE
DANA SCOTT
JOHN SHEPHERDSON

OXFORD LOGIC GUIDES

Recursion Theory for Metamathematics

RAYMOND M. SMULLYAN

Department of Philosophy
Indiana University

New York   Oxford
Oxford University Press
1993

Oxford University Press

Oxford   New York   Toronto
Delhi   Bombay   Calcutta   Madras   Karachi
Kuala Lumpur   Singapore   Hong Kong   Tokyo
Nairobi   Dar es Salaam   Cape Town
Melbourne   Auckland

and associated companies in
Berlin   Ibadan

Copyright © 1993 by Oxford University Press, Inc.

Published by Oxford University Press, Inc.,
200 Madison Avenue, New York, New York 10016

Oxford is a registered trademark of Oxford University Press

Library of Congress Cataloging-in-Publication Data
Smullyan, Raymond M.
Recursion theory for metamathematics / Raymond M. Smullyan.
p. cm.   (Oxford logic guides ; 22)
Includes bibliographical references and index.
ISBN 0-19-508232-X
1. Recursion theory. I. Title. II. Series.
QA9.6.S68   1993   511.3′5—dc20   92-40495

9 8 7 6 5 4 3 2 1

Printed in the United States of America
on acid-free paper

To Blanche

Preface

This volume, though a sequel to our book G.I.T. (Gödel's Incompleteness Theorems), can be read independently by those who have seen at least one proof of Gödel's incompleteness theorem for Peano Arithmetic (or at least know that the system is recursively axiomatizable). Our introductory chapter (Ch. 0) reviews all the background of G.I:T. (notations, definitions, key theorems and proofs) necessary for this volume.

Our study deals mainly with those aspects of recursion theory that have applications to the metamathematics of incompleteness, undecidability and related topics. It is both an introduction to the theory and a presentation of new results in the field.

The Gödel and Rosser incompleteness theorems were forerunners of many results of recursion theory—indeed, they were significantly responsible for opening up many portions of the field. But also, subsequent developments in pure recursion theory have shed further light on the phenomenon of incompleteness (and uniform incompleteness, as defined in our closing chapter). It is our purpose here to explore these fascinating interrelationships more deeply. Some related work of John Shepherdson (studied in G.I.T. and thoroughly reviewed here) plays a fundamental rôle in this study (particularly in Chapters 7 and 12).

Although we want this book to be thoroughly comprehensible to the reader with no prior knowledge of recursive function theory, we have written it just as much for the expert, who will find Chapters 4–12 (and particularly 6–12) to be the more mathematically original ones. A good deal of the latter portion of our *Theory of Formal Systems* [1961] is given an upgraded presentation here; we supply more motivation and make the results more easily accessible to the general reader, and several of the results and techniques are improved.

And now, for the reader familiar with the field, here is a very brief summary of what we do. Chapters 1–3 consist mainly of standard introductory material (basic closure properties of r.e. relations, the enumeration and iteration theorems, etc.) with Chapter 2 beginning an account of the author's earlier work on undecidability, essential undecidability and recursive inseparability. Chapters 4 and 5 lay the groundwork for Chapter 6 where we show (among other things) how the Ehrenfeucht-Feferman and Putnam-Smullyan theorems of 1960 can be proved without appeal to the recursion theorem or any other fixed point argument. The original proofs of these theorems leaned heavily on the use of creative sets and effectively inseparable pairs, but these turn out to be inessential. The really important underlying fact (Theorem B of Chapter 6) is that if (A, B) is a disjoint pair of r.e. (recursively enumerable) sets and is semi-doubly universal (i.e. for any disjoint r.e. sets C and D, there is a recursive function $f(x)$ which maps C into A and D into B), then (A, B) is doubly universal. We prove this key result in three different ways (each of which reveals certain facts not revealed by either of the others). Our first proof (Ch. 6) is along the lines of Smullyan [1963] and is based on the notion of *complete* effective inseparability as defined in Chapter 5. This proof requires no fixed point argument. Our third proof (Chapter 10) is pretty much the original one based on effective inseparability and uses the double recursion theorem (in the new form given in Chapter 9). Our second proof (Chapter 7) is of particular interest. We obtained it by taking the clever argument of Shepherdson [1961] (which we give in Ch. 0) and transfer it from the domain of formal theories to recursion theory itself. Now, we agree with Shepherdson's comment [1961] that his result (that every consistent axiomatizable Rosser system for binary relations is an exact Rosser system for sets) is apparently incomparable in strength with the Putnam-Smullyan result that every consistent axiomatizable Rosser system for sets in which all recursive functions of one argument are definable is an exact Rosser system for sets. But the interesting thing is that a slight modificaton of Shepherdson's argument (which we give in Chapter 7) yields a totally new proof of the Putnam-Smullyan theorem which uses virtually no recursion theory at all and furthermore yields a curious strengthening of the theorem.[1]

The next topic (Chapters 8 and 9) is recursion and multiple recursion theorems. Here we present a good deal of new material—far more than is needed for our applications to metamathematics. There are many subtle points involved and their interrelationships are quite fascinating. We divide recursion theorems into two types—*weak* and

---

[1] Many variations of Shepherdson's theorem appear in this volume. Some particularly interesting ones crop up in our final chapter.

strong. The former assert the existence of fixed points; the latter establish the existence of fixed point functions. The standard proof of the weak version of the recursion theorem requires only one application of the iteration theorem; the standard proof of the strong version requires two. We give a second proof of the strong recursion theorem that requires only one application of the iteration theorem. A minor modification of this proof yields an apparent strengthening of the recursion theorem that we call the *extended* recursion theorem. [Can it be derived as a corollary of any generally known recursion theorem? We do not know.] This extended recursion theorem provides alternative proofs of two special results of Chapter 9—the *symmetric* recursion theorem and a result we call "Theorem N". Both these results afford entirely new proofs of the author's double recursion theorem. Our original version of the double recursion theorem (given in T.F.S.) required a recursive pairing function $J(x,y)$ for its very statement; our new version does not and is accordingly more directly applicable to double productivity and effective inseparability.

These applications are given in Chapter 10. Chapter 11 is more for the specialist than the general reader and consists mainly in technical niceties concerning some strengthenings of earlier results.[2] Neither this chapter nor Chapter 10 is necessary for the final chapter (Ch. 12).

Our closing chapter is the principal one of the book and is written for the general reader and specialist alike. It introduces a variety of new concepts (e.g., effective Rosser systems, sentential and double sentential recursion properties, Rosser fixed point properties, uniform incompleteness) and ties them up with notions of earlier chapters. It contains attractive applications of recursion and double recursion theorems combined with Shepherdson type arguments which together yield the main results of this study.

I wish to thank Perry Smith for his many helpful corrections and suggestions.

---

[2] The specialist will probably find most interest in the section on feeble partial functions.

Contents

Recursion Theory for Metamathematics

Chapter 0

Prerequisites

As we remarked in the preface, although this volume is a sequel to our earlier volume G.I.T. (*Gödel's Incompleteness Theorems*), it can be read independently by those readers familiar with at least one proof of Gödel's first incompleteness theorem. In this chapter we give the notation, terminology and main results of G.I.T. that are needed for this volume. Readers familiar with G.I.T. can skip this chapter or perhaps glance through it briefly as a refresher.

I. Some General Incompleteness Theorems

§0. Preliminaries. We assume the reader to be familiar with the basic notions of first-order logic—the logical connectives, quantifiers, terms, formulas, free and bound occurrences of variables, the notion of *interpretations* (or models), truth under an interpretation, logical validity (truth under all interpretations), provability (in some complete system of first-order logic with identity) and its equivalence to logical validity (Gödel's completeness theorem). We let S be a system (theory) couched in the language of first-order logic with identity and with predicate and/or function symbols and with names for the natural numbers. A system S is usually presented by taking some standard axiomatization of first-order logic with identity and adding other axioms called the *non-logical* axioms of S.

We associate with each natural number n an expression $\bar{n}$ of S called the *numeral* designating n (or the *name* of n). We could, for example, take $\bar{0}, \bar{1}, \bar{2}, \ldots$, to be the expressions $0, 0', 0'', \ldots$, as we did in G.I.T. We have our individual variables arranged in some fixed

1

infinite sequence $v_1, v_2, \ldots, v_n, \ldots$[1] By $F(v_1, \ldots, v_n)$ we mean any formula whose free variables are all among $v_1, \ldots, v_n$, and for any (natural) numbers $k_1, \ldots, k_n$ by $F(\bar{k}_1, \ldots \bar{k}_n)$, we mean the result of substituting the numerals $\bar{k}_1, \ldots, \bar{k}_n$ for all free occurrences of $v_1, \ldots, v_n$ in F respectively. In particular, if $F(v_1)$ is a formula in which v_1 is the only free variable, then for any number n, $F(\bar{n})$ is the sentence resulting by substituting $\bar{n}$ for (all free occurrences of) v_1 in $F(v_1)$. [By a *sentence* we mean a formula in which there are no free variables.] And now, by $F[\bar{n}]$ (notice the square brackets!) we shall mean the sentence $\forall v_1(v_1 = \bar{n} \supset F(v_1))$. The sentences $F[\bar{n}]$ and $F(\bar{n})$ are logically equivalent. Hence, one is provable in $\mathcal{S}$ iff the other is—indeed the sentence $F(\bar{n}) \equiv F[\bar{n}]$ is a theorem of first-order logic with identity; hence it is provable in $\mathcal{S}$.[2] Actually, for *any* expression E of $\mathcal{S}$, whether a formula or not, we define $E[\bar{n}]$ to be the expression $\forall v_1(v_1 = \bar{n} \supset E)$. If E happens to be a formula, then so is $E[\bar{n}]$, but in any case, $E[\bar{n}]$ is well defined.

Gödel Numbering. We arrange all expressions of $\mathcal{S}$ (whether formulas or not) in some fixed 1–1 sequence $E_0, E_1, \ldots, E_n, \ldots$ and we call n the *Gödel number* of E_n.[3] For any expression X, we let $g(X)$ be its Gödel number (thus $g(E_n) = n$).

For any numbers a and b, by $r(a, b)$, we mean the Gödel number of $E_a[\bar{b}]$. The function $r(x, y)$ plays a crucial rôle and was referred to in G.I.T. as the *representation* function of $\mathcal{S}$. We let $d(x) = r(x, x)$. Thus, $d(n)$ is the Gödel number of $E_n[\bar{n}]$. We call $d(x)$ the *diagonal* function of $\mathcal{S}$.

We let P be the set of Gödel numbers of the provable formulas of $\mathcal{S}$, and R be the set of Gödel numbers of the refutable formulas of $\mathcal{S}$. [A formula is called *refutable* if its negation is provable.] For any set A of numbers, we let A^* be the set $d^{-1}(A)$—i.e. the set of all n such that $d(n) \in A$. Thus P^* is the set of all n such that $E_n[\bar{n}]$ is provable in $\mathcal{S}$, and R^* is the set of all n such that $E_n[\bar{n}]$ is refutable in $\mathcal{S}$. If E_n happens to be a formula, then $E_n[\bar{n}]$ is logically equivalent to $E_n(\bar{n})$. Hence, $n \in P^*$ iff (if and only if) $E_n(\bar{n})$ is provable, and so P^* is also the set of all n such that $E_n(\bar{n})$ is provable (in $\mathcal{S}$). Likewise, R^* is the set of all n such that $E_n(\bar{n})$ is refutable in $\mathcal{S}$.

---

[1] In G.I.T. we took these to be $(v_1), (v_{11}), \ldots$.

[2] The clever device of using $F[\bar{n}]$ instead of $F(\bar{n})$ to achieve diagonalization and self-reference is due to Alfred Tarski. It circumvents the necessity of arithmetizing substitution.

[3] A specific and handy Gödel numbering was given in Ch. 2, G.I.T.

By the *diagonalization* of an expression E_n, we mean $E_n[\bar{n}]$. Thus, $d(n)$ is the Gödel number of the diagonalization of E_n.

We call S *consistent* (sometimes *simply consistent*) if no sentence is both provable and refutable in S. We call S *complete* if every sentence is either provable or refutable in S; otherwise, S is called *incomplete*. A sentence X is called *undecidable* in S if it is neither provable nor refutable in S.

Representability. A formula $F(v_1, \ldots, v_n)$ is said to *represent* in S the set of all n-tuples $(a_1, \ldots, a_n)$ of numbers such that $F(\bar{a}_1, \ldots, \bar{a}_n)$ is provable in S. A relation $R(x_1, \ldots, x_n)$ (of natural numbers) is said to be *representable* in S if it is represented in S by some formula $F(v_1, \ldots, v_n)$.

We are regarding sets (of numbers) as special cases of relations (they are relations of one argument) and so the above definitions are also applicable to sets. Thus, $F(v_1)$ represents in S the set of all n such that $F(\bar{n})$ is provable in S. Thus, to say that $F(v_1)$ represents A is equivalent to the following condition: For all n, $F(\bar{n})$ is provable in $S \leftrightarrow n \in A$. [We write $\leftrightarrow$ to mean if and only if. Thus, $\leftrightarrow$ is a symbol of the metalanguage. For the equivalence symbol of the object language S, we use "$\equiv$".] Also, $F(v_1)$ represents the set of all n such that $F[\bar{n}]$ is provable in S (since $F(\bar{n})$ is provable in S iff $F[\bar{n}]$ is provable in S).

§1. Some Abstract Incompleteness Theorems.

A sentence X_n (with Gödel number n) will be called a *Gödel sentence* for a number set A (with respect to S understood) if the following condition holds:

$$X_n \text{ is provable in } S \leftrightarrow n \in A.$$

We shall call X_n a *negative* Gödel sentence for A if the following holds:

$$X_n \text{ is refutable in } S \text{ iff } n \in A.$$

Obviously every sentence is a Gödel sentence for the set P and is a negative Gödel sentence for the set R. If X is a Gödel sentence for R or if X is a negative Gödel sentence for P, then X is provable in S iff X is refutable in S; and if S is consistent, then X is undecidable in S.

Lemma A. *If A^* is representable in S, then there is a Gödel sentence for A and a negative Gödel sentence for A.*

More specifically, suppose $H(v_1)$ is a formula in v_1 and h is its Gödel number. Then

(1) If $H(v_1)$ represents A^*, then $H[\bar{h}]$ is a Gödel sentence for A.
(2) If the negation of $H(v_1)$ (i.e. $\sim H(v_1)$) represents A^*, then $H[\bar{h}]$ is a negative Gödel sentence for A.

Proof.

(1) Suppose $H(v_1)$ represents A^*. Then for any n, $H[\bar{n}]$ is provable iff $n \in A^*$. Hence, $H[\bar{h}]$ is provable iff $d(h) \in A$, but $d(h)$ is the Gödel number of $H[\bar{h}]$. Hence, $H[\bar{h}]$ is a Gödel sentence for A.
(2) Suppose $\sim H(v_1)$ represents A^*. In this case, $H[\bar{h}]$ is *refutable* in $\mathcal{S}$ iff $d(h) \in A$, and so $H[\bar{h}]$ is a negative Gödel sentence for A.

Remarks. Re (2), if A^* is represented in $\mathcal{S}$ by a formula $F(v_1)$, then $\sim F(v_1)$ is a formula whose negation represents A^* (since $\sim\sim F(v_1)$ is logically equivalent to $F(v_1)$). If A^* is representable in $\mathcal{S}$, then there is a formula $H(v_1)$ (namely $\sim F(v_1)$) whose negation represents A^*.

From (2) of Lemma A we have:

Theorem 1.[4] *Suppose $\mathcal{S}$ is consistent, and P^* is representable in $\mathcal{S}$. Then $\mathcal{S}$ is incomplete. More specifically, if $\mathcal{S}$ is consistent and $H(v_1)$ is a formula whose negation represents P^*, then $H(\bar{h})$ is undecidable in $\mathcal{S}$ where h is the Gödel number of $H(v_1)$.*

Proof. By (2) of Lemma A, the hypothesis implies that $H[\bar{h}]$ is a negative Gödel sentence for P. Hence, it is undecidable in $\mathcal{S}$ (if $\mathcal{S}$ is consistent), and $H(\bar{h})$ is also undecidable in $\mathcal{S}$.

In T.F.S. (Theory of Formal Systems) we proved the following "dual" of Theorem 1:

Theorem 1°.[5] *If $\mathcal{S}$ is consistent and if R^* is representable in $\mathcal{S}$, then $\mathcal{S}$ is incomplete. More specifically, suppose $\mathcal{S}$ is consistent, and $H(v_1)$ is a formula that represents R^* in $\mathcal{S}$. Then $H(\bar{h})$ is undecidable in $\mathcal{S}$ where h is the Gödel number of $H(v_1)$.*

---

[4] Th. 1 of Ch. 5, G.I.T.

[5] Th. 1°, Ch. 5, G.I.T

Proof. By (1) of Lemma A, the hypothesis implies that $H[\bar{h}]$ is a Gödel sentence for R. Hence it is undecidable in S (assuming consistency of S), and $H(\bar{h})$ is also undecidable in S.

Gödel's original incompleteness proof boils down to representing P^* in the system S under consideration. To do this, however, Gödel had to make a certain assumption about S—the assumption of "ω-consistency", to which we now turn.

§2. ω-Consistency.

We say that a system S is *ω-inconsistent* if there is a formula $F(v_1)$ such that the sentence $\exists v_1 F(v_1)$ is provable in S, yet all the sentences $F(\bar{0}), F(\bar{1}), \ldots, F(\bar{n}) \ldots$ are refutable in S. A system is called *ω-consistent* if it is not *ω-inconsistent*. When ω-consistency is under discussion, the phrase "simply consistent" is often used to mean *consistent* in order to avoid possible confusion. If a system S is (simply) inconsistent, then it is certainly ω-inconsistent since every formula is, then, provable. So an ω-consistent system is also simply consistent.

Enumerability in S. We say that a formula $F(v_1, v_2)$ *enumerates* a set A in S if for every number n, the following two conditions hold:

(1) If $n \in A$, then for some m, $F(\bar{n}, \bar{m})$ is provable in S.
(2) If $n \notin A$, then for every m, $F(\bar{n}, \bar{m})$ is refutable in S.

Suppose now that A is *enumerable* in S—enumerated in S by some formula $F(v_1, v_2)$. Then for any n, if $n \in A$, then for some m, the sentence $F(\bar{n}, \bar{m})$ is provable. Hence, by first-order logic, the sentence $\exists v_2 F(\bar{n}, v_2)$ is provable. Conversely, suppose $\exists v_2 F(\bar{n}, v_2)$ is provable. Does it follow that n is in A? Not necessarily, but if S is ω-consistent, then n is in A because if n were not in A, then all the sentences

$$F(\bar{n}, \bar{0}), F(\bar{n}, \bar{1}), \ldots, F(\bar{n}, \bar{m}) \ldots$$

would be refutable. Hence, S would be ω-inconsistent (since it is provable that $\exists v_2 F(\bar{n}, v_2)$). Thus, if S is ω-consistent and $F(v_1, v_2)$ enumerates A, then for every n, $n \in A$ iff $\exists v_2 F(\bar{n}, v_2)$ is provable. And so we have:

Lemma ω.[6] *Suppose S is ω-consistent. Then every set enumerable in S is representable in S—more specifically, if $F(v_1, v_2)$ enumerates*

---

[6] The ω-consistency lemma, Ch. 5, G.I.T.

A, then the formula $\exists v_2 F(v_1, v_2)$ represents A.

It follows from the above lemma and Theorem 1 that if P^* is *enumerable* in $\mathcal{S}$ and if $\mathcal{S}$ is ω-consistent, then $\mathcal{S}$ is incomplete. But more can be said.

Suppose $A(v_1, v_2)$ is a formula that enumerates the set P^*, and suppose that $\mathcal{S}$ is ω-consistent. Then by Lemma ω, the formula $\exists v_2 A(v_1, v_2)$ *represents* P^* in $\mathcal{S}$. Hence, the logically equivalent formula $\sim \forall v_2 \sim A(v_1, v_2)$ represents P^*. Therefore, the *negation* of the formula $\forall v_2 \sim A(v_1, v_2)$ represents P^*. So if h is the Gödel number of $\forall v_2 \sim A(v_1, v_2)$, then by Theorem 1, the sentence $\forall v_2 \sim A(\bar{h}, v_2)$—call this sentence G—is undecidable in $\mathcal{S}$. This means that G is neither provable nor refutable in $\mathcal{S}$. However, only the *simple* consistency of $\mathcal{S}$ is necessary to establish the unprovability of G because suppose G were provable, then h would be in the set P^*. Hence, for some number m, the sentence $A(\bar{h}, \bar{m})$ would be provable (since $A(v_1, v_2)$ enumerates P^*), and the sentence $\exists v_2 A(\bar{h}, v_2)$ would be provable which with the provability of $\forall v_2 \sim A(\bar{h}, v_2)$ (which is G) entails a *simple* inconsistency. Thus, if G is provable, then $\mathcal{S}$ is simply inconsistent. And so we have:

Theorem 2.[7] *Suppose that $A(v_1, v_2)$ enumerates P^* in $\mathcal{S}$, and h is the Gödel number of $\forall v_2 \sim A(v_1, v_2)$. Let G be the sentence*

$$\forall v_2 \sim A(\bar{h}, v_2).$$

Then

(1) *If $\mathcal{S}$ is simply consistent, then G is not provable in $\mathcal{S}$.*
(2) *If $\mathcal{S}$ is ω-consistent, then G is neither provable nor refutable in $\mathcal{S}$.*

Remarks. Theorem 2 constitutes an abstract form of Gödel's incompleteness theorem. For the type of system $\mathcal{S}$ investigated by Gödel, the set P^* (and also R^*) was shown to be *enumerable* in $\mathcal{S}$ without the assumption of ω-consistency. It was in passing from the enumerablity of P^* in $\mathcal{S}$ to the representability of P^* in $\mathcal{S}$ that ω-consistency entered the picture.

§3.   **Rosser's Method.**   J. Barkley Rosser obtained incompleteness, not by representing P^* in $\mathcal{S}$, but by representing some

---

[7]Th. 3, Ch. 5, G.I.T.

superset of P^* disjoint from R^*. [We call B a superset of A if A is a subset of B]. It turns out to be just as good if we represent some superset of R^* disjoint from P^* (as we will see).

Separability. Given two number sets A and B, we say that a formula $F(v_1)$ *weakly separates* A from B (in S) if $F(\bar{n})$ is provable when $n \in A$ and is not provable for $n \in B$—in other words, if $F(v_1)$ represents some superset A' of A disjoint from B. We say that $F(v_1)$ *strongly* separates A from B if $F(\bar{n})$ is provable for $n \in A$ and is refutable for $n \in B$. We say that A is weakly (strongly) separable from B (in S) if some formula $F(v_1)$ weakly (strongly) separates A from B. It is obvious that if S is consistent, then any formula that strongly separates A from B also weakly separates A from B. Also, if $F(v_1)$ represents A, then it weakly separates A from B whenever B is a set disjoint from A.

We proved (Theorem 1°) that if R^* is representable in S and S is (simply) consistent, then S is incomplete. The following theorem (on which Rosser's proof is based) is stronger:

Theorem 3.[8] *If R^* is weakly separable from P^* in S, then S is incomplete. More specifically, if $H(v_1)$ weakly separates R^* from P^* in S, then $H(\bar{h})$ is undecidable in S where h is the Gödel number of $H(v_1)$.*

Proof. Suppose $H(v_1)$ weakly separates R^* from P^* in S. Then for any number n, $H(\bar{n})$ is provable (in S) if $n \in R^*$, and $H(\bar{n})$ is not provable if $n \in P^*$. Hence, $H(\bar{h})$ is provable if $h \in R^*$ and is not provable if $h \in P^*$. Also, $h \in P^*$ if and only if $H(\bar{h})$ *is* provable. Thus if $H(\bar{h})$ is provable, then $h \in P^*$ and $H(\bar{h})$ is not provable, which is a contradiction. Hence, $H(\bar{h})$ is not provable. If $H(\bar{h})$ were refutable, then h would be in R^*. Hence, $H(\bar{h})$ would be provable, which it isn't. Therefore, $H(\bar{h})$ is not refutable either.

As a corollary we have:

Theorem 3.1. *If $H(v_1)$ strongly separates R^* from P^* and S is simply consistent, then $H(\bar{h})$ is undecidable in S where h is the Gödel number of $H(v_1)$.*

Remark. One can also show that if $H(v_1)$ is a formula whose negation weakly separates P^* from R^*, then $H(\bar{h})$ is undecidable in S.

---

[8] Th. 1, Ch. 6, G.I.T.

One also applies the notion of separability to relations of natural numbers. We say that $F(v_1, \ldots, v_n)$ *weakly separates* $R_1(x_1, \ldots, x_n)$ from $R_2(x_1, \ldots, x_n)$ if $H(\bar{a}_1, \ldots, \bar{a}_n)$ is provable for every n-tuple $(a_1, \ldots, a_n)$ in R_1 and for none in R_2. F is said to *strongly separate* R_1 from R_2 if F is provable for every n-tuple in R_1 and refutable for every n-tuple in R_2.

§3.1. Definability and Complete Representability. We regard sets (of natural numbers) as special cases of relations (sets are relations of one argument). We say that a formula $F(v_1, \ldots, v_n)$ *defines* a relation $R(x_1, \ldots, x_n)$ in $\mathcal{S}$ if F strongly separates R from its complement $\widetilde{R}$—in other words, if $F(\bar{a}_1, \ldots \bar{a}_n)$ is provable for every n-tuple $a_1, \ldots, a_n$ such that $R(a_1, \ldots, a_n)$ holds and is refutable in $\mathcal{S}$ for every n-tuple $(a_1, \ldots, a_n)$ such that $\sim R(a_1, \ldots, a_n)$ holds. We say that F *completely* represents R in $\mathcal{S}$ if F represents R, and its negation $\sim F$ represents the complement $\widetilde{R}$. For a consistent system $\mathcal{S}$, definability and complete representability are equivalent (as the reader can easily verify).

Let us note that if a formula $F(v_1, v_2)$ defines a relation $R(x, y)$ in $\mathcal{S}$ and if A is the domain of R (i.e. the set of all x such that $\exists y R(x, y)$), then $F(v_1, v_2)$ *enumerates* A in $\mathcal{S}$ (as the reader can easily verify). Thus, the domain of a binary relation definable in $\mathcal{S}$ is a set *enumerable* in $\mathcal{S}$. [If, also, $\mathcal{S}$ is ω-consistent, then the domain of a definable relation of $\mathcal{S}$ is *representable* in $\mathcal{S}$].

Exact Separation. We will later have need of the following notion: We say that a formula $F(v_1)$ *exactly* separates A from B in $\mathcal{S}$ if $F(v_1)$ represents A and $\sim F(v_1)$ represents B. This means that $F(\bar{n})$ is provable if $n \in A$, refutable if $n \in B$, and undecidable if n is in neither A nor B.

II. Arithmetic, Σ_0 and R.E. Relations

§4. Arithmetic, Σ_0 and R.E. Relations. We now consider formulas whose only function symbols are $+$ (plus), $\times$ (times) and $'$ (successor) and whose only predicate symbol (other than identity) is $\leq$ (less than or equal to). A formula $F(v_1, \ldots, v_n)$ is said to *express* the set of all n-tuples $(a_1, \ldots, a_n)$ such that $F(\bar{a}_1, \ldots, \bar{a}_n)$ is a *true* sentence (true, that is, under the standard interpretation). We call a formula $F(v_1, \ldots, v_n)$ *correct* if for every n-tuple $(a_1, \ldots, a_n)$,

the sentence $F(\bar{a}_1, \ldots, \bar{a}_n)$ is *true*. We let N be that first-order system whose axioms are all the correct formulas (this includes all logically valid formulas). Thus, the provable formulas of N are nothing more than the axioms of N. The system N is both consistent and complete (and also ω-consistent), but it is not recursively axiomatizable (as we shall define this). Of course, a formula expresses that relation which it represents (and also completely represents) in the complete system N. A relation (also a set) is called *arithmetic* if is representable in the complete system N.

Σ_0-**Relations.** For any variable v_i and any variable or numeral c, we write $(\forall v_i \leq c)(---)$ as an abbreviation of $\forall v_i(v_i \leq c \supset (---))$ and $(\exists v_i \leq c)$ as an abbreviation of $\exists v_i(v_i \leq c \wedge (---))$. We refer to $(\forall v_i \leq c)$ and $(\exists v_i \leq c)$ as *bounded* quantifiers.

By an *atomic* Σ_0-formula, we mean a formula of any of the four forms

$$c_1 + c_2 = c_3, \quad c_1 \times c_2 = c_3, \quad c_1 = c_2, \quad c_1 \leq c_2,$$

where each c_1 and c_2 is a variable or a numeral. We then define the class of Σ_0-formulas by the following inductive rules:

(1) Every atomic Σ_0-formula is a Σ_0-formula.
(2) If F_1 and G_1 are Σ_0-formulas, then so are

$$F_1 \wedge G_1, \quad F_1 \vee G_1, \quad F_1 \supset G_1, \quad F_1 \equiv G_1, \quad \sim F_1.$$

(3) If F is a Σ_0-formula, v_i is a variable and c is a numeral or a variable *distinct* from v_i, then $(\forall v_i \leq c)F$ and $(\exists v_i \leq c)F$ are Σ_0-formulas.

Thus, in a Σ_0-formula, all the quantifiers are bounded. Σ_0-formulas are also called *constructive arithmetic*. Given a Σ_0-sentence (*i.e.* a Σ_0-formula with no free variables), one can effectively decide whether it is true or false.

By a Σ_0-relation, (also called a constructive arithmetic relation) we mean a relation expressed by some Σ_0-formula.

Σ_1-**Relations.** By a Σ_1-formula, we mean one of the form $\exists v_i F$, where F is a Σ_0-formula. Thus, a Σ_1-formula is the *unbounded* existential quantification of a Σ_0-formula. And by a Σ_1-relation (or an r.e. (recursively enumerable)) relation, we mean a relation expressed by a Σ_1-formula.[9]

---

[9] Many equivalent definitions of *recursively enumerable* are to be found in the literature. The Σ_1-definition fits in best with the purposes of this volume and is the one we

In G.I.T. we gave a proof of the following well-known fact.

Theorem 4.[10]

(a) *Every Σ_0-relation is also Σ_1.*

(b) *If $R(x_1, \ldots, x_n, y)$ is a Σ_1-relation, then so is the relation $\exists y R(x_1, \ldots, x_n, y)$.*

(c) *The union and intersection of two Σ_1-relations are Σ_1-relations.*

(d) *If $R(x_1, \ldots, x_n, y, z)$ is Σ_1, then so is the relation $(\exists y \leq z) R(x_1, \ldots, x_n, y, z)$.*

(e) *If $R(x_1, \ldots, x_n, y, z)$ is Σ_1, then so is the relation $(\forall y \leq z) R(x_1, \ldots, x_n, y, z)$.*

Complete details of the proof can be found in G.I.T. The only tricky case is (e), and the idea is this. Suppose $R(x_1, \ldots, x_n, y, z)$ is Σ_1. Then it is of the form $\exists w S(x_1, \ldots, x_n, y, z, w)$, where S is a Σ_0-relation, and so $(\forall y \leq z) R(x_1, \ldots, x_n, y, z)$ is the relation

$$(\forall y \leq z) \exists w S(x_1, \ldots, x_n, y, z, w).$$

If this relation holds, then for every z and every $y \leq z$, there is a number w_y such that $S(x_1, \ldots, x_n, y, z, w_y)$. If we let v be the maximum of the numbers $w_0, \ldots, w_z$, we can rewrite the relation as

$$\exists v (\forall y \leq z)(\exists w \leq v) S(x_1, \ldots, x_n, y, z, w),$$

which is Σ_1.

Recursive Relations. We define a set or relation to be *recursive* if it and its complement are both r.e. A function $f(x_1, \ldots, x_n)$ is called recursive if the relation $f(x_1, \ldots, x_n) = y$ is recursive. Let us note that if the relation $f(x_1, \ldots, x_n) = y$ is recursively enumerable, then it must be recursive since the complementary relation $f(x_1, \ldots, x_n) \neq y$ can be written

$$\exists z (f(x_1, \ldots, x_n) = z \wedge z \neq y).$$

Hence, it is r.e. (as can be seen using Theorem 4).

Σ_0-Complete Systems. We call $\mathcal{S}$ Σ_0-*complete* if all true Σ_0-sentences are provable in $\mathcal{S}$. This condition implies that all r.e. sets are *enumerable* in $\mathcal{S}$. Suppose $\mathcal{S}$ is Σ_0-complete and A is an r.e. set. Then A is the domain of a Σ_0-relation $R(x_1, x_2)$, and this relation is

---

accordingly adopt.

[10] Proposition C, Ch. 4, G.I.T.

expressed by some Σ_0-formula $F(v_1, v_2)$. Then $F(v_1, v_2)$ *enumerates* the set A in S (because if $n \in A$, then $R(n, m)$ holds for some m. Hence the true Σ_0-sentence $F(\bar{n}, \bar{m})$ is provable in S; whereas if $n \notin A$, then for every m, $\sim R(n, m)$ holds. Hence for every m, the true Σ_0-sentence $\sim F(\bar{n}, \bar{m})$ is provable in S).

§5. Axiomatizable Systems.

So far we have made no assumptions about the Gödel numbering of the expressions of S. Now it will be necessary to do so.

For any expressions X and Y, by XY we mean X followed by Y. We let $x * y$ be the Gödel number of $E_x E_y$. For any number n, we let $\text{num}(n)$ be the Gödel number of the numeral $\bar{n}$. We now define a Gödel numbering y to be *acceptable* if the functions $x * y$ and $\text{num}(x)$ are recursive functions. We, henceforth, assume that our Gödel numbering is acceptable.[11]

We now define S to be *recursively axiomatizable* (*axiomatizable*, for short) if the set P of Gödel numbers of the provable formulas of S is recursively enumerable. [This definition is actually independent of which acceptable Gödel numbering is chosen, and the definition can be given without reference to any Gödel numbering by using either the canonical language of Post, the elementary formal systems of Smullyan, or the algorithms of Markov.]

Now, assuming the Gödel numbering is acceptable, it is easily verified that the function $r(x, y)$ (the Gödel number of $E_x[\bar{y}]$—which is the Gödel number of $\forall v_1(v_1 = \bar{y} \supset E_x)$) is a recursive function of x and y[12] and, hence, the diagonal function $d(x)$ (which is $r(x, x)$) is recursive. It easily follows that for any r.e. set A, the set A^* (i.e. $d^{-1}(A)$) is r.e. Thus if S is axiomatizable, then the set P is r.e. Hence the set P^* is also r.e. If P is r.e., then so is R and, hence, so is R^*. Thus if S is axiomatizable, then the sets P^* and R^* are both r.e.

Suppose now that S is an ω-consistent axiomatizable system in which all true Σ_0-sentences are provable. Then all r.e. sets are enumerable in S. Hence they are representable in S (by the ω-consistency lemma). Also P^* is r.e. (since S is axiomatizable). Hence S must be incomplete by Theorem 1, and so we have the following generalization of Gödel's theorem:

---

[11] Examples of acceptable (and handy) Gödel numberings were given in Ch. 2, G.I.T.

[12] cf. Ch. 2, G.I.T. for details

Theorem 5.[13]   *If S is an axiomatizable ω-consistent system in which all true Σ_0-sentences are provable, then S is incomplete.*

§6.   Rosser Systems.

We call S a *Rosser system for sets* if for any r.e. sets A and B, the set $A-B$ is strongly separable from $B-A$ in S. More generally, for any positive integer n, we say that S is a *Rosser system for n-ary relations* if for any r.e. relations $R_1(x_1, \ldots, x_n)$ and $R_2(x_1, \ldots, x_n)$, the relation R_1-R_2 (that is to say $R_1 \wedge \sim R_2$) is strongly separable from R_2-R_1. We call S a *Rosser system* if it is a Rosser system for sets and for relations of any number of arguments.

If S is a Rosser system for sets, then obviously for any *disjoint* r.e. sets A and B, the set A is strongly separable from B in S (since then $A - B = A$ and $B - A = B$). The converse also happens to hold (as we will later see).

Suppose now that S is a *simply* consistent axiomatizable Rosser system for sets. Then the sets P^* and R^* are both r.e., and by consistency, they are disjoint. R^* is then strongly separable from P^* in S, and by Theorem 3.1, S is then incomplete. And so we have:

Theorem 6. *If S is a simply consistent axiomatizable Rosser system for sets, then S is incomplete.*

§7.   The Systems P.A., (Q) and (R).

We call a system S_1 a *subsystem* of S_2, and we say that S_2 is an *extension* of S_1 if all the provable formulas of S_1 are provable in S_2. We now consider some significant axiomatizable subsystems of N—the system P.A. (Peano Arithmetic), the subsystem (Q) of P.A. (a variant of one due to Raphael Robinson) and the further subsystem (R) (also due to Robinson).

We start with the system (Q) (which will play a key role in this volume). It has only finitely many non-logical axioms—namely the following nine. [We are using $'$ for the successor function.]

N_1: $v_1' = v_2' \supset v_1 = v_2$
N_2: $\sim v_1' = \bar{0}$
N_3: $v_1 + \bar{0} = v_1$
N_4: $v_1 + v_2' = (v_1 + v_2)'$

---

[13] Th. A, Ch. 5, G.I.T.

N_5: $v_1 \times \bar{0} = \bar{0}$
N_6: $v_1 \times v_2' = (v_1 \times v_2) + v_1$
N_7: $v_1 \leq \bar{0} \equiv v_1 = \bar{0}$
N_8: $v_1 \leq v_2' \equiv (v_1 \leq v_2 \lor v_1 = v_2')$
N_9: $v_1 \leq v_2 \lor v_2 \leq v_1$

The System P.A. The non-logical axioms of P.A. are infinite in number. They consist of the nine axioms of (Q) together with the infinitely many axioms of scheme N_{10} below—one for each formula $F(v_1)$ (which may contain other free variables than v_1). These are the so-called *induction* axioms.

N_{10}: $[F(\bar{0}) \land \forall v_1(F(v_1) \supset F(v_1'))] \supset \forall v_1 F(v_1)$

The system P.A. appears, on the surface, to be a complete system, but it turns out not to be (as we will see).

The System (R). This system also has infinitely many non-logical axioms—all formulas of one of the following five forms (m and n are any numbers).

Ω_1: $\bar{m} + \bar{n} = \bar{k}$, where $m + n = k$
Ω_2: $\bar{m} \times \bar{n} = \bar{k}$, where $m \times n = k$
Ω_3: $\bar{m} \neq \bar{n}$, where $m \neq n$
Ω_4: $v_1 \leq \bar{n} \equiv (v_1 = \bar{0} \lor \ldots \lor v_1 = \bar{n})$
Ω_5: $v_1 \leq \bar{n} \lor \bar{n} \leq v_1$

One importance of the system (R) is that it is Σ_0-complete—indeed the system (R_0) consisting of Ω_1–Ω_4 is Σ_0-complete. We call a Σ_0-sentence *correctly* decidable in (R_0) if it is either true and provable in (R_0) or false and refutable in (R_0). An easy induction on degrees of Σ_0-sentences (*i.e.* number of occurrences of logical connectives and quantifiers) shows that all Σ_0-sentences are correctly decidable in (R_0)[14]. Hence all true Σ_0-sentences are provable in (R_0) (and hence in (R)).

The next important fact about (R) is that it is a subsystem of (Q) (and hence of P.A.), which makes (Q) (and P.A.) Σ_0-complete. Briefly, the proof is this: Using N_3 and N_4, one shows by mathematical induction on m that all sentences of axiom scheme Ω_1 are provable in (Q). Having established this, one uses N_5 and N_6, and by mathematical induction on m, one shows that all sentences of Ω_2 are provable in (Q). As for Ω_3, we use N_1 and N_2 and show by mathematical induction on k that for all *positive* numbers p, the sentence

---

[14] see Part II, Chapter 5, G.I.T. for details

$\bar{k} \neq \overline{k+p}$ (and hence also $\overline{k+p} \neq \bar{k}$) is provable in (Q). As for Ω_4 using N_7 and N_8, one shows by mathematical induction on n that $v_1 \leq \bar{n} \equiv (v_1 = \bar{0} \vee \ldots \vee v_1 \leq \bar{n})$ is provable in (Q). At this point, we see that (R_0) is a subsystem of (Q) (in fact, of the subsystem (Q_0) of (Q) consisting of just N_1–N_8). As for Ω_5, this is immediate from N_9, and so (R) is a subsystem of (Q).[15] We thus have:

Theorem 7. *The system (R) is a subsystem of (Q) which, in turn, is a subsystem of P.A. These systems are all Σ_0-complete (in fact, so are (R_0) and (Q_0)).*

The Axiomatizability of P.A. In G.I.T., Chapters 3 and 4, we proved the axiomatizability of P.A. A minor modification of the proof yields the axiomatizablity of the systems (Q) and (R). The proof is lengthy (although simpler, we believe, than the standard proofs) and will not be repeated here.

§8. More on Rosser Systems.

The following theorem is of fundamental importance:

Theorem 8. *The systems (R), (Q) and P.A. are Rosser systems.*

This follows from the Σ_0-completeness of these systems and from lemma S below. We call a system $\mathcal{S}$ an extension of Ω_4 and Ω_5 if all instances of Ω_4 and Ω_5 are provable in $\mathcal{S}$.

Lemma S.[16] Suppose $\mathcal{S}$ is an extension of Ω_4 and Ω_5. Then for any relations $R_1(x_1, \ldots, x_n)$ and $R_2(x_1, \ldots, x_n)$, if R_1 and R_2 are enumerable in $\mathcal{S}$, then $R_1 - R_2$ is strongly separable from $R_2 - R_1$ in $\mathcal{S}$. More specifically, if $F_1(x_1, \ldots, x_n, y)$ and $F_2(x_1, \ldots, x_n, y)$ are formulas that respectively enumerate the relation $R_1(x_1, \ldots, x_n)$ and $R_2(x_1, \ldots, x_n)$, then the formula

$$\exists y [F_1(x_1, \ldots, x_n, y) \wedge (\forall z \leq y) \sim F_2(x_1, \ldots, x_n, z)]$$

strongly separates $R_1 - R_2$ from $R_2 - R_1$. [Alternately,

$$\forall y [F_1(x_1, \ldots, x_n, y) \supset (\exists z \leq y) F_2(x_1, \ldots, x_n, z)]$$

strongly separates $R_2 - R_1$ from $R_1 - R_2$.]

Lemma S is an easy consequence of the following lemma

---

[15] Details of the proof can be found in the proofs of Proposition 3 and 4, Chapter 5, G.I.T.

[16] Separation lemma, Ch. 6, G.I.T.

Lemma S_0. Given two formulas $F_1(y)$ and $F_2(y)$, let X be the sentence $\exists y(F_1(y) \wedge (\forall z \leq y) \sim F_2(z))$. Suppose S is an extension of Ω_4 and Ω_5. Then for any number n:

(1) If $F_1(\bar{n})$ is provable and if $F_2(\bar{0}), \ldots, F_2(\bar{n})$ are all refutable, then X is provable.
(2) If $F_2(\bar{n})$ is provable and if $F_1(\bar{0}), \ldots, F_1(\bar{n})$ are all refutable, then X is refutable.

Proof of Lemma S_0. Suppose all instances of Ω_4 and Ω_5 are provable in S.

1. Suppose $F_1(\bar{n})$ is provable and $F_2(\bar{0}), \ldots, F_2(\bar{n})$ are all refutable. From the latter it easily follows by Ω_4 that $(\forall z \leq \bar{n}) \sim F_2(z)$ is provable. Hence $F_1(\bar{n}) \wedge (\forall z \leq \bar{n}) \sim F_2(z)$ is provable, and by first-order logic, $\exists y(F_1(y) \wedge (\forall z \leq y) \sim F_2(z))$ is provable. Thus X is provable.
2. Suppose $F_2(\bar{n})$ is provable and $F_1(\bar{0}), \ldots, F_1(\bar{n})$ are all refutable. Then by $\Omega_4, (\forall y \leq \bar{n}) \sim F_1(y)$ is provable. Hence the open formula $y \leq \bar{n} \supset \sim F_1(y)$ is provable, and $F_1(y) \supset \sim (y \leq \bar{n})$ is provable. By Ω_5, $F_1(y) \supset \bar{n} \leq y$ is provable.

$$F_1(y) \supset (\bar{n} \leq y \wedge F_2(\bar{n}))$$

is provable (since $F_2(\bar{n})$ is provable),

$$F_1(y) \supset (\exists z \leq y) F_2(z)$$

is provable,

$$\sim (F_1(y) \wedge (\forall z \leq y) \sim F_2(z))$$

is provable, and

$$\sim \exists y(F_1(y) \wedge (\forall z \leq y) \sim F_2(z))$$

is provable; and X is refutable.

From Theorem 8 and Theorem 6 we have:

Theorem 8.1. *Every simply consistent axiomatizable extension of (R) is incomplete.*

The above theorem has sometimes been referred to as the "Gödel-Rosser incompleteness theorem".

§9. The Non-axiomatizability of N; Tarski's Theorem.

We shall call a system S *arithmetic* if the set of Gödel numbers of the provable formulas of S is an arithmetic set. Since every r.e. set is obviously arithmetic, then every axiomatizable system is also an arithmetic system. And now we have the following result of Alfred Tarski:

Theorem 9. *The system N is not even arithmetic let alone axiomatizable.*

Since N is complete, Theorem 9 follows from the following result:

Theorem $9^\sharp$. *Every arithmetic subsystem of N is incomplete.*

Proof of Th. $9^\sharp$. Suppose S is an arithmetic subsystem of N. Then the set P of Gödel numbers of the provable formulas of S is arithmetic. Hence its complement $\widetilde{P}$ is arithmetic, and the set $\widetilde{P}^*$ is arithmetic. [For any arithmetic set A, the set A^* is arithmetic since $A^* = d^{-1}(A)$, and $d(x)$ is a recursive function. Hence the relation $d(x) = y$ is arithmetic, and $d^{-1}(A)$ is thus the set of all n satisfying the arithmetic condition $\exists y(d(n) = y \wedge y \in A)$.] Since the set $\widetilde{P}^*$ is arithmetic, it is represented in N by some formula $H(v_1)$. Then by Lemma A, $H[\bar{h}]$ is a Gödel sentence for $\widetilde{P}$ with respect to N, *i.e.* $H[\bar{h}]$ is *true* iff its Gödel number is in $\widetilde{P}$. Thus $H[\bar{h}]$ is true iff it is not provable in S. Thus the sentence is either true and not provable in S or false and provable in S. The latter alternative is ruled out by the assumption that S is a subsystem of N. Thus $H[\bar{h}]$ is true but not provable in S. Since $H[\bar{h}]$ is true, then $\sim H[\bar{h}]$ is false and not provable in S, and $H[\bar{h}]$ is an undecidable sentence of S.

Three Incompleteness Proofs for Peano Arithmetic. We now have three different methods of showing the incompleteness of Peano Arithmetic.

    I. Assuming that all provable sentences of P.A. are true, P.A. is then a subsystem of N. Since P.A. is also axiomatizable, it is incomplete by Theorem 9.

This proof is the simplest of all (and apparently the least well known). It doesn't involve proving that P.A. is a Rosser system, nor that P.A. is Σ_0-complete. This is the first proof we gave in G.I.T. However, this proof involves the strongest metamathematical assumption of all, namely that all provable sentences of P.A. are true.

    II. Under the weaker metamathematical assumption that P.A. is ω-consistent, we have Gödel's original method: Since P.A. is axiomatizable and Σ_0-complete, (assuming ω-consistency) it is then incomplete by Theorem 5.

This proof does involve showing that P.A. is Σ_0-complete (which we did by showing it to be an extension of (R)).

III. Under the still weaker assumption that P.A. is *simply* consistent, since P.A. is an axiomatizable extension of (R), it is incomplete by Theorem 8.1. This is Rosser's proof.

Other incompleteness proofs for P.A. will follow from results proved in the present volume.

§10.   More on Separation.

Consider two r.e. sets A and B. Then there are Σ_0-relations, $R_1(x,y)$ and $R_2(x,y)$, such that A is the domain of R_1 and B is the domain of R_2. Let A' be the set of all n such that

$$\exists y(R_1(n,y) \wedge (\forall z \leq y) \sim R_2(n,z)),$$

and let B' be the set of all n such that

$$\exists y(R_2(n,y) \wedge (\forall z \leq y) \sim R_1(n,z)).$$

Let us say that y *puts* n in A iff $R_1(n,y)$ and that y puts n in B iff $R_2(n,y)$. Then $n \in A$ iff some y puts $n \in A$, and $n \in B$ iff some y puts n in B. Let us say that n is put in A *before* n is put in B if there is some y that puts n in A and no $z \leq y$ puts n in B. The set A' is then the set of all n such that n is put in A before it is put in B; B' is the set of all n such that n is put in B before it is put in A.

The sets A' and B' are clearly disjoint, and $A - B \subseteq A'$ and $B - A \subseteq B'$. Also, the sets A' and B' are easily seen to be r.e. (this follows from the various parts of Theorem 4). And so we have:

Theorem 10.[17]   *For any two r.e. sets A and B, there are disjoint r.e. sets A' and B' such that $A - B \subseteq A'$ and $B - A \subseteq B'$.*

Suppose now that $\mathcal{S}$ is a system such that for any two disjoint r.e. sets A and B, the set A is strongly separable from B in $\mathcal{S}$. Now suppose A and B are r.e. sets, not necessarily disjoint. Then by Th. 10, there are disjoint r.e. sets A' and B' such that $A - B \subseteq A'$ and $B - A \subseteq B'$. There is then a formula $F(v_1)$ which strongly separates A' from B' in $\mathcal{S}$, and it is obvious that $F(v_1)$ strongly separates $A - B$ from $B - A$ in $\mathcal{S}$. And so we have:

---

[17] Th. 5, Ch. 6, G.I.T.

Theorem 10.1.[18] *If every disjoint pair of r.e. sets is strongly separable in S, then S is a Rosser system for sets.*

Of course, Theorems 10 and 10.1 also hold for n-ary relations where $n > 1$. The reader can easily verify this.

§11.   Definability of Functions in S.

The notion of a function being definable in S is defined differently by different authors. We will say that a formula $F(v_1, \ldots, v_n, v_{n+1})$ weakly defines a function $f(x_1, \ldots, x_n)$ if it defines the relation $f(x_1, \ldots, x_n) = y$. We will say that $F(v_1, \ldots, v_n, v_{n+1})$ *strongly* defines, or more briefly, defines $f(x_1, \ldots, x_n)$ in S if, in addition, for any numbers $a_1, \ldots, a_n, b$ such that $f(a_1, \ldots, a_n) = b$, the sentence

$$\forall v_{n+1}(F(\bar{a}_1, \ldots, \bar{a}_n, v_{n+1}) \equiv v_{n+1} = \bar{b})$$

is provable in S.

In G.I.T. (Lemma to Theorem 3 of §3, Ch. 8) we proved:

Lemma. *If S is an extension of Ω_4 and Ω_5, then every function weakly definable in S is strongly definable in S.*

The idea behind the proof is that if S is an extension of Ω_4 and Ω_5 and if $F(v_1, v_2)$ is a formula that weakly defines the function $f(x)$ in S, then the formula

$$F(v_1, v_2) \wedge \forall v_3(F(v_1, v_3) \supset v_2 \leq v_3)$$

will strongly define $f(x)$ in S.

The reader should be able to prove this as an exercise (or consult G.I.T. for details).

If S is a Rosser system, then it is obvious that all *recursive* relations are definable in S. Hence all recursive functions are weakly definable in S. If also S is an extension of Ω_4 and Ω_5, then by the above lemma, all recursive functions are strongly definable in S. Since (R) is a Rosser system and is also an extension of Ω_4 and Ω_5, we thus have:

Theorem 11.[19] *All recursive functions are strongly definable in (R) (and, hence, in every extension of (R), in particular in the systems (Q) and P.A.).*

---

[18] Corollary of Th. 5, Ch. 6, G.I.T.

[19] Th. 3, Ch. 8, G.I.T.

Admissible Functions. We call a function $f(x)$ *admissible* in S if for every formula $G(v_1)$, there is a formula $H(v_1)$ such that for every number n, the sentence $H(\bar{n}) \equiv G(\overline{f(n)})$ is provable in S. Now suppose $f(x)$ is strongly definable in S by a formula $F(v_1, v_2)$. Then, given a formula $G(v_1)$, if we take $H(v_1)$ to be the formula

$$\exists v_2(F(v_1, v_2) \wedge G(v_2)),$$

it is not difficult to verify that for any n, the sentence

$$H(\bar{n}) \equiv G(\overline{f(n)})$$

is provable in S[20] and so we have:

Theorem 11.1.[21] *If $f(x)$ is strongly definable in S, then $f(x)$ is admissible in S.*

Corollary. *All recursive functions of one argument are admissible in (R) (and hence in every extension of (R)).*

The significance of admissibility is this: Suppose that for all n, $H(\bar{n}) \equiv G(\overline{f(n)})$ is provable in S. Then if A is the set represented in S by $G(v_1)$, and B is the set represented by $\sim G(v_1)$, then $H(v_1)$ clearly represents $f^{-1}(A)$, and $\sim H(v_1)$ represents $f^{-1}(B)$. Then by Theorem 11.1 we have:

Theorem 11.2.[22] *Suppose $f(x)$ is strongly definable in S (or even admissible in S). Then*

(1) *If A is representable in S, then so is $f^{-1}(A)$.*
(2) *If (A, B) is exactly separable in S, then so is $(f^{-1}(A), f^{-1}(B))$.*
(3) *If A is definable in S, then so is $f^{-1}(A)$.*

III.   Shepherdson's Theorems

We now turn to some results of John Shepherdson that we considered in G.I.T. and which will play an important role in this volume.

At the time of Gödel's proof, the only known way of showing that all r.e. sets are representable in P.A. involved the assumption of ω-consistency. Well, in 1960, A. Ehrenfeucht and S. Feferman

---

[20] cf. Th. 2, Ch. 8, G.I.T. for details

[21] Th. 2, Ch. 8, G.I.T.

[22] Corollary of Th. 2, Ch. 8, G.I.T.

showed that all r.e. sets are representable in every *simply* consistent axiomatizable extension of the system (R). In fact their proof showed the following stronger result:

Theorem E.F. *If S is any simply consistent axiomatizable Rosser system for sets in which all recursive functions of one argument are strongly definable, then all r.e. sets are representable in S.*

Their proof combined a Rosser-type argument with a celebrated result of John Myhill which we will study in this volume.

We call S an *exact* Rosser system for sets if every disjoint pair of r.e. sets is *exactly* separable in S. In 1960, H. Putnam and R. Smullyan proved the following strengthening of the Ehrenfeucht-Feferman theorem.

Theorem P.S. *If S is a consistent axiomatizable Rosser system for sets in which all recursive functions of one argument are strongly definable, then S is an exact Rosser system for sets.*

Of course Theorem P.S. shows that every consistent axiomatizable extension of (R) is an exact Rosser system for sets. The proof of Th. P.S. involved a "double analogue" of Myhill's theorem that we will study in this volume.

Now, in 1961, Shepherdson gave a remarkably direct proof that every axiomatizable consistent extension of (R) is an exact Rosser system for sets. He proved the following two theorems:

Theorem S_1—Shepherdson's Representation Theorem.   *If S is a consistent axiomatizable Rosser system for binary relations, then every r.e. set is representable in S.*

Theorem S_2—Shepherdson's Exact Separation Theorem. *Under the same hypothesis, S is an exact Rosser system for sets.*

The hypothesis of the Shepherdson theorems is apparently incomparable in strength with the hypothesis of the E.F. and P.S. theorems. Both yield alternative proofs that every consistent axiomatizable extension of (R) is an exact Rosser system for sets.

We now turn to the proofs of the Shepherdson theorems (and shall in fact prove some slightly stronger results that will be important for this volume).

§12.   Shepherdson's Representation Lemma.   For any expression E and any numbers m and n, we let $E[\bar{m}, \bar{n}]$ be the ex-

pression $\forall v_2(v_2 = \bar{m} \supset \forall v_1(v_1 = \bar{n} \supset E))$. If E is a formula in which v_1 and v_2 are the only free variables, then $E[\bar{m}, \bar{n}]$ is a sentence logically equivalent to $E(\bar{m}, \bar{n})$, and so $E[\bar{m}, \bar{n}]$ is then provable in S iff $E(\bar{m}, \bar{n})$ is provable.

We now define the relation $P^\sharp$ to be the set of all ordered pairs (m, n) such that $E_n(\bar{m}, \bar{n})$ is provable in S, and $R^\sharp$ to be the set of all (m, n) such that $E_n(\bar{m}, \bar{n})$ is refutable in S. It is easily verified that if S is axiomatizable, then $P^\sharp$ and $R^\sharp$ are both r.e. relations.

Lemma 1—Shepherdson's Representation Lemma. *For any set A, if the relation $x \in A \wedge \sim P^\sharp(x, y)$ is weakly separable in S from the relation $P^\sharp(x, y) \wedge x \notin A$, then A is representable in S.*

We will, in fact, need the following stronger lemma (for purposes of this volume).

Lemma 1*. *For any relation $R(x, y)$, if $E_h(v_1, v_2)$ is a formula that weakly separates in S the relation $R(x, y) \wedge \sim P^\sharp(x, y)$ from $P^\sharp(x, y) \wedge \sim R(x, y)$, then for every number n: $R(n, h)$ iff $E_h(\bar{n}, \bar{h})$ is provable in S.*

Note. Lemma 1 follows from Lemma 1* by taking $R(x, y)$ iff $x \in A$ (regardless of y).

Proof of Lemma 1*. Assume hypothesis. Then for all n and m:

(1) $(R(n, m) \wedge \sim P^\sharp(n, m)) \supset E_h(\bar{n}, \bar{m})$ is provable (in S)
(2) $(P^\sharp(n, m) \wedge \sim R(n, m)) \supset E_h(\bar{n}, \bar{m})$ is not provable.

Taking h for m we have:

(1') $(R(n, h) \wedge \sim P^\sharp(n, h)) \supset E_h(\bar{n}, \bar{h})$ is provable
$\qquad\qquad\qquad\qquad\qquad\qquad \supset P^\sharp(n, h)$.
(2') $(P^\sharp(n, h) \wedge \sim R(n, h)) \supset E_h(\bar{n}, \bar{h})$ not provable
$\qquad\qquad\qquad\qquad\qquad\qquad \supset \sim P^\sharp(n, h)$.

From (1)', it follows that $R(n, h) \supset P^\sharp(n, h)$. From (2)' , it follows that $P^\sharp(n, h) \supset R(n, h)$. Therefore, $R(n, h)$ iff $P^\sharp(n, h)$, and so $R(n, h)$ iff $E_h(\bar{n}, \bar{h})$ is provable in S.

Lemma 1 easily yields Theorem S_1, and, in fact, Lemma 1* yields the following stronger result:

Theorem S_1^*. *Suppose S is a consistent axiomatizable Rosser system for binary relations. Then for any r.e. relation $R(x, y)$, there is a formula $E_h(v_1, v_2)$ such that for all n:*

$$E_h(\bar{n}, \bar{h}) \text{ is provable in } S \leftrightarrow R(n, h).$$

§13.   Shepherdson's Exact Separation Lemma.

Lemma 2—Shepherdson's Exact Separation Lemma. *For any disjoint sets A and B, if the relation*

$$(x \in A \vee R^{\sharp}(x, y)) \wedge \sim (x \in B \vee P^{\sharp}(x, y))$$

is strongly separable in S from the relation

$$(x \in B \vee P^{\sharp}(x, y)) \wedge \sim (x \in A \vee R^{\sharp}(x, y)),$$

and if S is consistent, then A is exactly separable from B in S.

We will need the following stronger lemma:

Lemma 2*. *For any disjoint relations $M_1(x, y)$ and $M_2(x, y)$, if*

$$(M_1 \cup R^{\sharp}) - (M_1 \cup P^{\sharp})$$

is strongly separated in S from

$$(M_2 \cup P^{\sharp}) - (M_1 \cup R^{\sharp})$$

by a formula $E_h(v_1, v_2)$, and if S is consistent, then for any number n:

(a) $E_h(\bar{n}, \bar{h})$ *is provable* $\leftrightarrow M_1(n, h)$
(b) $E_h(\bar{n}, \bar{h})$ *is refutable* $\leftrightarrow M_2(n, h)$.

Proof. Assume hypothesis. Then for all n:

(1) $[(M_1(n, h) \vee R^{\sharp}(n, h)) \wedge \sim (M_2(n, h) \vee P^{\sharp}(n, h))] \supset E_h(\bar{n}, \bar{h})$ is provable $\supset P^{\sharp}(n, h)$
(2) $[(M_2(n, h) \vee P^{\sharp}(n, h)) \wedge \sim (M_1(n, h) \vee R^{\sharp}(n, h))] \supset E_h(\bar{n}, \bar{h})$ is refutable $\supset R^{\sharp}(n, h)$

Since M_1 and M_2 are assumed disjoint and $R^{\sharp}$ and $P^{\sharp}$ are disjoint (by our assumption that S is consistent), conclusions (a) and (b) follow from (1) and (2) by propositional logic.

From Lemma 2* we easily get the following strengthening of Theorem S_2:

Theorem S_2^*. *Suppose S is a consistent axiomatizable Rosser system for binary relations. Then for any disjoint r.e. relations $R_1(x, y)$ and $R_2(x, y)$, there is a formula $E_h(v_1, v_2)$ such that for all n:*

(a) $E_h(\bar{n}, \bar{h})$ *is provable in* $\mathcal{S} \leftrightarrow R_1(n, h)$
(b) $E_h(\bar{n}, \bar{h})$ *is refutable in* $\mathcal{S} \leftrightarrow R_2(n, h)$.

Exercise. Let $R(x, y)$ be an r.e. relation and for any number n, let R_n be the set of all numbers x such that $R(n, x)$. Prove that for any consistent axiomatizable Rosser system for binary relations, there is a number h such that R_h is represented in the system by a formula whose Gödel number is h. [Hint: Use Lemma 1*.]

Chapter I

Recursive Enumerability and Recursivity

Having proved that Peano Arithmetic is incomplete, we can ask another question about the system. Is there any algorithm (mechanical procedure) by which we can determine which sentences are provable in the system and which are not? This brings us to the subject of recursive function theory, to which we now turn.

We are defining a relation (or set) to be r.e. (recursively enumerable) iff it is Σ_1, and to be recursive iff it and its complement are r.e. An equivalent definition of recursive enumerability is representability in some finitely axiomatizable system (as we will prove). Many other characterizations of recursive enumerability and recursivity can be found in the literature (cf., e.g., Kleene [1952], Turing [1936], Post [1944], Smullyan [1961], Markov [1961]), but the Σ_1-characterization fits in best with the overall plan of this volume. The fact that so many different and independently formulated definitions turn out to be equivalent adds support to a thesis proposed by Church—namely that any function that is effectively calculable in the intuitive sense is a recursive function. Interesting discussions of Church's thesis can be found in Kleene [1952] and Rogers [1967].

In this chapter, we establish a few basic properties of recursive enumerability that will be needed in just about all the chapters that follow.

I. Some Basic Closure Properties

§1. Some Closure Properties. It will be convenient to regard sets as special cases of relations (sets are thus relations of one argument or relations of *degree 1*).

It will be convenient to use the λ-notation "$\lambda x_1, \ldots, x_n : (\ldots)$", read "the set of all n-tuples $(x_1, \ldots, x_n)$ such that $(\ldots)$". For example, for any relation $R(x_1, x_2, x_3)$, the relation $\lambda x_1 x_2 x_3 : R(x_2, x_2, x_1)$ is the set of all triples (x_1, x_2, x_3) (of natural numbers) such that $R(x_2, x_2, x_1)$ holds. We sometimes write "$x : (\ldots)$" for "$\lambda x : (\ldots)$".

Explicit Definability. We say that a relation $S(x_1, \ldots, x_n)$ is *explicitly* definable from a relation $R(x_1, \ldots, x_k)$ if $S(x_1, \ldots, x_n)$ can be written in the form

$$\lambda x_1, \ldots, x_n : R(\xi_1, \ldots, \xi_k),$$

where each ξ_i is either one of the variables $x_1, \ldots, x_k$ or is a natural number.

For example, if R is a relation of degree 3, and S is the set of all quintuples $(x_1, x_2, x_3, x_4, x_5)$ such that $R(x_4, x_1, 7)$, then S is explicitly definable from R (since $S(x_1, x_2, x_3, x_4, x_5)$ can be written as $\lambda x_1, x_2, x_3, x_4, x_5 : R(x_4, x_1, 7)$).

Suppose S is explicitly definable from R, and R is Σ_1. Then it is obvious that S is also Σ_1. For the above example, if $F(v_1, v_2, v_3)$ is a Σ_1-formula expressing the relation $R(x_1, x_2, x_3)$, then the relation $S(x_1, x_2, x_3, x_4, x_5)$ is expressed by the Σ_1-formula

$$F(v_4, v_1, \bar{7}) \wedge v_2 = v_2 \wedge v_3 = v_3 \wedge v_5 = v_5.$$

It is also true that any relation explicitly definable from a Σ_0-relation is Σ_0, and any relation explicitly definable from an arithmetic relation is arithmetic.

We say that a class $\mathcal{C}$ of relations is closed under explicit definability if for every $R \in \mathcal{C}$, all relations explicitly definable from R are also in $\mathcal{C}$.

We would like to remark that the notion of *explicit definability* can be defined without use of λ-notation as follows. For every positive n and every positive $i \leq n$, let P_i^n be the function of n arguments defined by the condition

$$P_i^n(x_1, \ldots, x_n) = x_i.$$

We call these functions P_i^n *projection functions* (they are sometimes called *identity* functions). For every positive n and any number a, let C_a^n be the function of n arguments defined by the condition

$$C_a^n(x_1, \ldots, x_n) = a.$$

The functions C_a^n (for various n and a) are called *constant functions*. Then we can say that a relation $S(x_1, \ldots, x_k)$ is explicitly definable from $R(x_1, \ldots, x_n)$ if there are functions

$$f_1(x_1, \ldots, x_k), \ldots, f_n(x_1, \ldots, x_k),$$

each of which is either a projection function or a constant function, such that for all $x_1, \ldots, x_k$:

$$S(x_1, \ldots, x_k) \leftrightarrow R(f_1(x_1, \ldots, x_k), \ldots, f_n(x_1, \ldots, x_k)).$$

For example, suppose $S = \lambda x_1 x_2 x_3 x_4 : R(x_3, 7, x_2)$. Then

$$S(x_1, x_2, x_3, x_4) \leftrightarrow R(P_3^4(x_1, x_2, x_3, x_4), C_7^4(x_1, x_2, x_3, x_4),$$
$$P_2^4(x_1, x_2, x_3, x_4)).$$

Unions and Intersections. For any two relations $R_1(x_1, \ldots, x_n)$ and $R_2(x_1, \ldots, x_n)$, by $R_1 \cup R_2$ (the *union* of R_1 and R_2), we mean

$$\lambda x_1, \ldots, x_n : (R_1(x_1, \ldots, x_n) \vee R_2(x_1, \ldots, x_n))$$

and by the *intersection* $R_1 \cap R_2$ of R_1, R_2, we mean

$$\lambda x_1, \ldots, x_n : (R_1(x_1, \ldots, x_n) \wedge R_2(x_1, \ldots, x_n)).$$

We know from Th. 4, Ch. 0 that if R_1 and R_2 are both r.e., then so are the relations $R_1 \cup R_2$ and $R_1 \cap R_2$.

Quantifications. For any relation $R(x_1, \ldots, x_n, y)$ by its *existential* quantification, we mean

$$\lambda x_1, \ldots, x_n : \exists y \, R(x_1, \ldots, x_n, y)$$

and by its *universal* quantification, we mean

$$\lambda x_1, \ldots, x_n : \forall y \, R(x_1, \ldots, x_n, y).$$

We know from Chapter 0 that the existential quantification of an r.e. relation is r.e. (because existential quantifications of Σ-relations are Σ, and Σ-relations are the same as Σ_1-relations). The universal quantification of an r.e. relation is in general *not* r.e. (as we will see).

Finite Quantifications. By the *finite* existential quantification of a relation $R(x_1, \ldots, x_n, y, z)$, we mean

$$\lambda x_1, \ldots, x_n, y : (\exists z \leq y) R(x_1, \ldots, x_n, y, z).$$

By the finite universal quantification of R, we mean

$$\lambda x_1, \ldots, x_n, y : (\forall z \leq y) R(x_1, \ldots, x_n, y, z).$$

By Th. 4, Ch. 0, if R is r.e., then the finite existential quantification of R and the finite universal quantification of R are both r.e.

Collecting together the facts we now know, we summarize them as:

Proposition A. *The class C of r.e. relations is closed under explicit definability, unions, intersections, existential quantifications, and finite universal (and existential) quantifications and contains the relations $x + y = z$, $x \cdot y = z$, $x = y$ and $x \neq y$.*

Exercise 1. (a) Prove that if C is any class of relations having the closure properties above, then C must contain all r.e. relations. [Hint: Use induction on degrees of Σ-formulas].

    (b) Using (a) and Proposition A, show that a relation is r.e. if and only if it belongs to every class C having the above closure properties. [This provides a purely set-theoretic characterization of r.e. relations; this characterization is not stated with reference to the metamathematical notions of *formula* and *truth*].

Exercise 2. Give similar set-theoretic characterizations of Σ_0-relations and arithmetic relations.

§2. Recursive Relations.

We recall that we are calling a relation *recursive* iff it and its complement are r.e.

Proposition B. *The class of recursive relations is closed under complementation, union, intersection, finite quantifications, and explicit definability, and it contains the relations $x + y = z$, $x \cdot y = z$, $x = y$ and $x \neq y$.*

Proof.

1. Suppose R is recursive. Then R and $\tilde{R}$ are both r.e. and $\tilde{R}$ and $\tilde{\tilde{R}}$ are both r.e. (since $\tilde{\tilde{R}} = R$); so $\tilde{R}$ is recursive.
2. Suppose R_1 and R_2 are recursive relations of the same degree. Then R_1, $\tilde{R_1}$, R_2 and $\tilde{R_2}$ are all r.e., and $R_1 \cup R_2$ is r.e. by Proposition A. Its complement is $\tilde{R_1} \cap \tilde{R_2}$, which is r.e. by Proposition A. Thus $R_1 \cup R_2$ is recursive, and $R_1 \cap R_2$ is r.e. by Proposition A. Its complement is $\tilde{R_1} \cup \tilde{R_2}$, which is r.e. by Proposition A. Hence $R_1 \cap R_2$ is recursive.
3. Suppose $R(x_1, \ldots, x_n, y, z)$ is recursive. Since it is r.e., by Proposition A, the relation $(\exists z \leq y) R(x_1, \ldots, x_n, y, z)$ is r.e. Since the relation $\tilde{R}(x_1, \ldots, x_n, y, z)$ is r.e., then by Proposi-

tion A, so is the relation $(\forall z \leq y)\tilde{R}(x_1, \ldots, x_n, y, z)$, but this is the complement of the relation $(\exists z \leq y)R(x_1, \ldots, x_n, y, z)$. Therefore the relation $(\exists z \leq y)R(x_1, \ldots, x_n, y, z)$ is recursive. The proof that the relation $(\forall z \leq y)R(x_1, \ldots, x_n, y, z)$ is recursive is similar.

4. Suppose $R(x_1, \ldots, x_k)$ is recursive, and S is the relation $\lambda x_1, \ldots, x_n$: $R(\xi_1, \ldots, \xi_k)$ (each ξ_i is either one of the variables $x_1, \ldots, x_n$ or a constant). Since R is r.e., so is S (by Proposition A.) Also $\tilde{S}$ is the relation $\lambda x_1, \ldots, x_n : \tilde{R}(\xi_1, \ldots, \xi_k)$, and since $\tilde{R}$ is r.e., so is $\tilde{S}$. Thus S is recursive.

5. Since the *functions* $x + y$ and $x \cdot y$ and the identity function are r.e., then they are recursive.

§3. Some Consequences.

Proposition 1. *Suppose* $f_1(x_1, \ldots, x_n), \ldots, f_k(x_1, \ldots, x_n)$ *are recursive. Let* $R(x_1, \ldots, x_k)$ *be any relation, and let* $S(x_1, \ldots, x_n)$ *be the relation*

$$R(f_1(x_1, \ldots, x_n), \ldots, f_k(x_1, \ldots, x_n)).$$

(1) *If* R *is r.e., then so is* S.
(2) *If* R *is recursive, then so is* S.

Proof. $S(x_1, \ldots, x_n)$ holds iff

$$\exists z_1 \ldots \exists z_k (f_1(x_1, \ldots, x_n) = z_1 \wedge \ldots \wedge f_k(x_1, \ldots, x_n) = z_k \\ \wedge R(z_1, \ldots, z_k)).$$

If R is r.e., it then easily follows from Proposition A that S is r.e. Also $\tilde{S}(x_1, \ldots, x_n)$ iff

$$\exists z_1 \ldots \exists z_k (f_1(x_1, \ldots, x_n) = z_1 \wedge \ldots \wedge f_k(x_1, \ldots, x_n) = z_k \\ \wedge \tilde{R}(z_1, \ldots, z_k)),$$

so if $\tilde{R}$ is r.e., then so is $\tilde{S}$. Therefore, if R is recursive, then so is S.

Proposition 2. (a) *For any recursive functions*

$$f_1(x_1, \ldots, x_n), \ldots, f_k(x_1, \ldots, x_n) \text{ and } g(x_1, \ldots, x_k),$$

the function $g(f_1(x_1, \ldots, x_n), \ldots, f_k(x_1, \ldots, x_n))$ *is recursive.*
(b) *For any recursive function* $f(x_1, \ldots, x_n)$ *and any r.e. set* A, *the relation* $f(x_1, \ldots, x_n) \in A$ *is r.e. If* A *is recursive, then*

the above relation is recursive.

(c) *For any recursive function $f(y)$ and any r.e. relation $R(x_1, \ldots, x_n, y)$, the relation $R(x_1, \ldots, x_n, f(y))$ is r.e. If R is recursive, then so is the relation $R(x_1, \ldots, x_n, f(y))$.*

(d) *For any r.e. relation $R(x_1, \ldots, x_n)$ and any recursive function $f(x_1, \ldots, x_n)$, the set of all numbers $f(x_1, \ldots, x_n)$ such that $R(x_1, \ldots, x_n)$ is an r.e. set.*

Proof. (a) follows from (2) of Proposition 1 taking for R the recursive relation $g(x_1, \ldots, x_k) = y$.

(b) follows from Proposition 1, taking R to be the set A.

(c) follows from Proposition 1, since $R(x_1, \ldots, x_n, f(y))$ can be written as

$$R(I(x_1), \ldots, I(x_n), f(y)),$$

where $I(x)$ is the identity function (which is obviously recursive).

As for (d), let A be the set of all numbers $f(x_1, \ldots, x_n)$ such that $R(x_1, \ldots, x_n)$. Then for any number x, we have

$$x \in A \leftrightarrow \exists x_1 \ldots \exists x_n(R(x_1, \ldots, x_n) \wedge x = f(x_1, \ldots, x_n)).$$

Proposition 3. *If $R(x_1, \ldots, x_n, y)$ is r.e. (recursive), then the relations*

$$(\exists z \leq y)R(x_1, \ldots, x_n, z) \text{ and } (\forall z \leq y)R(x_1, \ldots, x_n, z)$$

are r.e. (recursive).

Proof. Given a relation $R(x_1, \ldots, x_n, y)$, let S be the set of all $(n+2)$-tuples $(x_1, \ldots, x_n, y, z)$ such that $R(x_1, \ldots, x_n, y)$. Then S is explicitly definable from R and

$$(\exists z \leq y)R(x_1, \ldots, x_n, z) \leftrightarrow (\exists z \leq y)S(x_1, \ldots, x_n, y, z).$$

If R is r.e. (recursive), then so is S. Hence, so is

$$(\exists z \leq y)S(x_1, \ldots, x_n, y, z)$$

and

$$(\exists z \leq y)R(x_1, \ldots, x_n, z).$$

The proof for $(\forall z \leq y)R(x_1, \ldots, x_n, z)$ is similar.

Proposition 4. *If $R(x_1, \ldots, x_n, y)$ is r.e. (recursive), then so are the relations*

$$(\exists z < y)R(x_1, \ldots, x_n, z) \text{ and } (\forall z < y)R(x_1, \ldots, x_n, z).$$

Proof.

(1) $(\exists z < y)R(x_1,\ldots,x_n,z) \leftrightarrow (\exists z \leq y)(R(x_1,\ldots,x_n,z) \wedge z < y).$

(2) $\begin{aligned}(\forall z < y)R(x_1,\ldots,x_n,z) &\leftrightarrow (\forall z \leq y)(z < y \supset R(x_1,\ldots,x_n,z))\\ &\leftrightarrow (\forall z \leq y)(z = y \vee R(x_1,\ldots,x_n,z))\end{aligned}$

The rest of the proof is easily obtained by using Propositions A and B.

Proposition 5. *If $R(x_1,\ldots,x_n,y,z)$ is recursive, then for any recursive function $f(x)$, the relation*

$$(\exists z \leq f(y))R(x_1,\ldots,x_n,y,z)$$

is recursive.

Proof. Define $S(x_1,\ldots,x_n,y)$ iff $(\exists z \leq f(y))R(x_1,\ldots,x_n,y,z)$. Then:

(1) $S(x_1,\ldots,x_n,y) \leftrightarrow \exists w(w = f(y) \wedge (\exists z \leq w)R(x_1,\ldots,x_n,y,z),$
(2) $\tilde{S}(x_1,\ldots,x_n,y) \leftrightarrow \exists w(w = f(y) \wedge (\forall z \leq w) \sim R(x_1,\ldots,x_n,y,z).$

From (1) it easily follows (using earlier propositions) that if R is r.e., then so is S. From (2), it follows that if $\tilde{R}$ is r.e., then so is $\tilde{S}$. Therefore, if R is recursive, then so is S.

Exercise 3. Show that if $R(x_1,\ldots,x_n,y,z)$ is recursive, then for any recursive function $f(x)$, the relation

$$(\forall z \leq f(y))R(x_1,\ldots,x_n,y,z)$$

is recursive.

Regular Relations. We shall call a relation $R(x_1,\ldots,x_n,y)$ *regular* if for every $x_1,\ldots,x_n$ there is at least one y such that $R(x_1,\ldots,x_n,y)$. For a regular relation $R(x_1,\ldots,x_n,y)$, by $\mu y R(x_1,\ldots,x_n,y)$, we mean the smallest number y such that $R(x_1,\ldots,x_n,y)$.

Proposition 6. *If $R(x_1,\ldots,x_n,y)$ is recursive and regular, then the function $\mu y R(x_1,\ldots,x_n,y)$ (as a function of $x_1,\ldots,x_n$) is recursive.*

Proof. Suppose R is recursive and regular. Let

$$f(x_1,\ldots,x_n) = \mu y R(x_1,\ldots,x_n,y).$$

Then for all $x_1,\ldots,x_n$ and y:

$$f(x_1,\ldots,x_n) = y \leftrightarrow [R(x_1,\ldots,x_n,y) \wedge (\forall z < y) \sim R(x_1,\ldots,x_n,z)].$$

Since R and $\widetilde{R}$ are both r.e., then the relation

$$R(x_1, \ldots, x_n, y) \wedge (\forall z < y) \sim R(x_1, \ldots, x_n, z)$$

must be r.e. (Why?) Hence $f(x_1, \ldots, x_n)$ is a recursive function.

Proposition 7.

(1) *For any r.e. relation $R(x_1, \ldots, x_n, y_1, \ldots, y_k)$ and any numbers $a_1, \ldots, a_k$, the relation $R(x_1, \ldots, x_n, a_1, \ldots, a_k)$ (as a relation among $x_1, \ldots, x_n$) is r.e. [In λ-notation, this relation is written: $\lambda x_1 \ldots x_n : R(x_1, \ldots, x_n, a_1, \ldots, a_k)$.]*

(2) *For any recursive function $f(x_1, \ldots, x_n, y_1, \ldots, y_k)$ and numbers $a_1, \ldots, a_k$, the function $f(x_1, \ldots, x_n, a_1, \ldots, a_k)$ is recursive. Also for any numbers $b_1, \ldots, b_n$, the function $f(b_1, \ldots, b_n, y_1, \ldots, y_k)$ (as a function of $y_1, \ldots, y_k$) is recursive.*

(3) *For any r.e. relation $R(x, y)$, its inverse $R(y, x)$ [In λ-notation, the inverse is written: $\lambda x, y : Ry, x$] is r.e.*

(4) *For any recursive function $f(x, y)$, the function $f(x, x)$ is recursive.*

Proof. In each case, the new relation (or function) is explicitly definable from the old one.

II. Recursive Pairing Functions

§4. Recursive Pairing Functions.

By a *recursive pairing function*, we mean a 1-1 recursive function $J(x, y)$ such that there exist recursive functions $K(x)$ and $L(x)$ such that for all x,

$$J(Kx, Lx) = x.$$

We recall that a function $f(x, y)$ is called 1-1 if for all numbers x_1, y_1, x_2 and y_2, if $f(x_1, y_1) = f(x_2, y_2)$, then $x_1 = x_2$ and $y_1 = y_2$.

There are many ways to construct recursive pairing functions. The standard method is to use Georg Cantor's enumeration of all ordered pairs of natural numbers, which is this: We first take all ordered pairs (x, y) whose sum is 0 (there is only one such pair, viz. $(0, 0)$). Then we take all pairs (x, y) whose sum is 1; there are only two such pairs, and we take them in the order $(0, 1), (1, 0)$. Then we take all ordered pairs whose sum is 3 in the order $(0, 3), (1, 2), (2, 1), (3, 0)$, and so forth. Thus $(0, 0)$ is the 0th term of our enumeration, and for any

n, if (x, y) is the nth term, then the $(n + 1)$th term is

$$(x + 1, y - 1), \quad \text{if} \quad y \neq 0, \text{ and}$$
$$(0, x + 1), \quad \text{if} \quad y = 0.$$

The first 13 terms of the enumeration are $(0,0)$, $(0,1)$, $(1,0)$, $(0,2)$, $(1,1)$, $(2,0)$, $(0,3)$, $(1,2)$, $(2,1)$, $(3,0)$, $(0,4)$, $(1,3)$, $(2,2)$, $(3,1)$, (4.0). We then define $J(x, y)$ to be that number n such that (x, y) is the nth element of the enumeration.

One can show that

$$J(x, y) = \frac{1}{2}(x + y)(x + y + 1) + x,$$

(see Exercise 4 below), and so the function $J(x, y)$ is recursive. Note that

$$J(x, y) = z \leftrightarrow (x + y)(x + y + 1) + 2x = 2z.$$

Since each number x is $J(x_1, x_2)$ for exactly one pair (x_1, x_2), we define $K(x) = x_1$ and $L(x) = x_2$. Then $J(Kx, Lx) = J(x_1, x_2) = x$, and so $J(Kx, Lx) = x$.

Also, for any numbers x and y, if we let $z = J(x, y)$, then $Kz = x$ and $Lz = y$—in other words, $KJ(x, y) = x$ and $LJ(x, y) = y$.

Since $J(x, y)$ is recursive, so are the functions Kx and Lx because

$$Kx = y \leftrightarrow (\exists z \leq x)J(y, z) = x,$$

and

$$Lx = y \leftrightarrow (\exists z \leq x)J(z, y) = x.$$

We thus have:

Proposition 8. *There is a 1-1 recursive function $J(x, y)$ and recursive functions Kx and Lx such that for all numbers x and y:*

(1) $J(Kx, Lx) = x$,
(2) $KJ(x, y) = x$ and $LJ(x, y) = y$.

The functions Kx and Lx are called the *inverse* functions of $J(x, y)$.

Corollary. *For any recursive function $f(x, y)$, there is a recursive function $\phi(x)$ such that for all $x, y : f(x, y) = \phi J(x, y)$.*

Proof. Let $\phi(x) = f(Kx, Lx)$. Then for all x and y:

$$\phi J(x, y) = f(KJ(x, y), LJ(x, y)) = f(x, y).$$

Exercise 4. We use the well known algebraic fact that for any number n, the sum $0 + 1 + \cdots + n = \frac{1}{2}n(n+1)$.

(a) In terms of n, how many ordered pairs (a, b) are there such that $a + b = n$?

(b) In terms of n, how many ordered pairs (a, b) are there such that $a + b < n$?

(c) Given an ordered pair (x, y), state in terms of x, how many ordered pairs (a, b) there are such that $a + b = x + y$ and $a < x$.

(d) Using (a), (b) and (c), show that

$$J(x, y) = \frac{1}{2}(x + y)(x + y + 1) + x.$$

Solution.

(a) Obviously there are $n + 1$ such pairs (viz. $(0, n)$, $(1, n - 1)$, ..., $(n, 0)$).

(b) Since for each $m < n$, there are $m + 1$ ordered pairs (a, b) such that $a + b = m$, and there are $0 + 1 + 2 + \cdots + n$ ordered pairs (a, b) such that $a + b < n$. So the answer is $\frac{1}{2}n(n+1)$.

(c) The answer is obviously x (the ordered pairs are $(0, x + y)$, $(1, x + y - 1)$, ..., $(x - 1, y + 1)$).

(d) Let $n = x + y$. The number $J(x, y)$ is the number of ordered pairs which precede (x, y) in the Cantor enumeration. By (b), the number of such pairs whose sum is less than n is $\frac{1}{2}n(n+1)$. By (c) the number of such pairs whose sum is n is x. Therefore the number of such pairs all told is $\frac{1}{2}n(n + 1) + x$, which is $\frac{1}{2}(x + y)(x + y + 1) + x$.

§5. **The Functions** $J_n(x_1, \ldots, x_n)$. For each $n \geq 2$, we define the function $J_n(x_1, \ldots, x_n)$ by the following inductive scheme:

$$J_2(x_1, x_2) = J(x_1, x_2);$$
$$J_3(x_1, x_2, x_3) = J(J_2(x_1, x_2), x_3);$$
$$\vdots$$
$$J_{n+1}(x_1, \ldots, x_n, x_{n+1}) = J(J_n(x_1, \ldots, x_n), x_{n+1}).$$

An obvious induction argument shows that for each $n \geq 2$, the function $J_n(x_1, \ldots, x_n)$ is a 1-1 recursive function whose range is the set N of natural numbers.

Proposition 9.

(1) *For any r.e. set A and any $n \geq 2$, the relation*

$$J_n(x_1, \ldots, x_n) \in A$$

   is r.e.

(2) *For any r.e. relation $R(x_1, \ldots, x_n)$, the set of all numbers $J_n(x_1, \ldots, x_n)$ such that $R(x_1, \ldots, x_n)$ is an r.e. set.*

Proof. By (*b*) and (*d*) of Proposition 2.

Proposition 9 largely enables us to reduce the theory of r.e. relations to the theory of r.e. sets.

The next proposition will have a nice application in Chapter 3.

Proposition 10. *For any positive integers m and n and any r.e. relation $R(x_1, \ldots, x_m, y_1, \ldots, y_n)$ of $m+n$ arguments, there is an r.e. relation $M(x, y)$ such that for all numbers $x_1, \ldots, x_m$ and $y_1, \ldots, y_n$:*

$$R(x_1, \ldots, x_m, y_1, \ldots, y_n) \leftrightarrow M(J_m(x_1, \ldots, x_m), J_n(y_1, \ldots, y_n)).$$

Proof. Define $M(x, y)$ to hold iff there are numbers $x_1, \ldots, x_m$ and $y_1, \ldots, y_n$ such that $x = J_m(x_1, \ldots, x_m)$, $y = J_n(y_1, \ldots, y_n)$ and $R(x_1, \ldots, x_m, y_1, \ldots, y_n)$. Since the functions J_m and J_n are 1-1, then it is immediate that

$$M(J_m(x_1, \ldots, x_m), J_n(y_1, \ldots, y_n)) \leftrightarrow R(x_1, \ldots, x_m, y_1, \ldots, y_n).$$

Also, the relation $M(x, y)$ is r.e., since it can be written as

$$\exists x_1 \ldots \exists x_m \exists y_1 \ldots \exists y_n (x = J_m(x_1, \ldots, x_m) \wedge y = J_n(y_1, \ldots, y_n) \wedge$$

$$R(x_1, \ldots, x_m, y_1, \ldots, y_n)).$$

The Functions K_i^n. For each $n \geq 2$, we define the recursive functions $K_1^n(x), K_2^n(x), \ldots, K_n^n(x)$ by the following inductive scheme on $n \geq 2$. For $n = 2$, we let $K_1^2(x) = Kx$ and $K_2^2(x) = Lx$. Now suppose $n \geq 2$ and the functions $K_1^n, \ldots, K_n^n$ are defined. Then we let $K_i^{n+1}(x) = K_i^n(Kx)$, for $i \leq n$, and we let $K_{n+1}^{n+1}(x) = L(x)$. [For example, $K_1^3(x) = KKx$, $K_2^3(x) = LKx$ and $K_3^3(x) = Lx$.] An obvious induction on $n \geq 2$ yields:

Proposition 11. *For any $n \geq 2$, any n-tuple $(x_1, \ldots, x_n)$, and each $i \leq n$,*

$$K_i^n J_n(x_1, \ldots, x_n) = x_i.$$

Exercise 5. Prove Proposition 11.

Exercise 6. Using Proposition 11, prove that for any $n \geq 2$ and any x, $J_n(K_1^n(x), \ldots, K_n^n(x)) = x$.

Proposition 12. *For any $n \geq 2$ and any numbers $i_1, \ldots, i_n$, each $\leq n$, there is a recursive function $g(x)$ such that for every n-tuple $(x_1, \ldots, x_n)$, we have*

$$g J_n(x_1, \ldots, x_n) = J_n(x_{i_1}, \ldots, x_{i_n}).$$

Proof. Take $g(x)$ to be $J(K_{i_1}^n x, \ldots, K_{i_n}^n x)$. Then use Proposition 11.

Exercise 7. In terms of the functions J, K and L, give an explicit description of a recursive function $g(x)$ such that for any quadruple (x_1, x_2, x_3, x_4),

$$g J_4(x_1, x_2, x_3, x_4) = J_4(x_1, x_2, x_4, x_3).$$

III. *Representability and Recursive Enumerability*

§6.   We now turn to the study of first order systems. We recall that a system S is called *axiomatizable* if the set P of Gödel numbers of its provable formulas is recursively enumerable.

Theorem 1. *If S is axiomatizable, then every set and relation representable in S is r.e.*

We shall first prove Theorem 1 for sets. We use the function $r(x, y)$ of Chapter 0 ($r(x, y)$ is the Gödel number of $E_x[\bar{y}]$). We know that the relation $r(x, y) = z$ is Σ_1; hence the function $r(x, y)$ is recursive.

Now, suppose S is axiomatizable. This means that the set P of Gödel numbers of the provable formulas of S is r.e. Suppose A is a set representable in S. Then some formula $H(v_1)$ represents A in S. Let h be the Gödel number of $H(v_1)$. Then for any n, $n \in A \leftrightarrow H[\bar{n}]$ is provable in $S \leftrightarrow r(h, n) \in P$. Since $r(x, y)$ is a recursive function, so is $r(h, x)$ (as a function of x), by (2) of Proposition 7. Since A is the inverse image of P under this function and P is r.e., then A is r.e. by (b) of Prop. 2. This proves that every set representable in S is r.e.

To prove this for relations of degree ≥ 2, we first define $E[\bar{a}_1, \ldots, \bar{a}_n]$ to be the expression

$$\forall v_1 \ldots \forall v_n((v_1 = \bar{a}_1 \wedge \cdots \wedge v_n = \bar{a}_n) \supset E).$$

[We could alternatively take the expression

$$\forall v_1(v_1 = \bar{a}_1 \supset (\forall v_2(v_2 = \bar{a}_2 \supset \cdots \supset (\forall v_n(v_n = \bar{a}_n \supset E))))).$$

to be $E[\bar{a}_1, \ldots, \bar{a}_n]$.] Under either definition, $E[\bar{a}_1, \ldots, \bar{a}_n]$ is provable in $\mathcal{S}$ iff $E(\bar{a}_1, \ldots, \bar{a}_n)$ is provable in $\mathcal{S}$. This conclusion holds because $E[\bar{a}_1, \ldots, \bar{a}_n] \equiv E(\bar{a}_1, \ldots, \bar{a}_n)$ is logically valid. For each $n \geq 1$ we define $r_n(x, y_1, \ldots, y_n)$ to be the Gödel number of $E_x[\bar{y}_1, \ldots, \bar{y}_n]$. [For $n = 1$, $r_1(x, y) = r(x, y)$.] We leave it to the reader to verify that for each $n > 1$, the function $r_n(x, y_1, \ldots, y_n)$ is recursive. Now if a relation $R(x_1, \ldots, x_n)$ is represented in $\mathcal{S}$ by a formula $H(v_1, \ldots, v_n)$ with Gödel number h, then

$$R = \lambda x_1 \ldots x_n : r_n(h, x_1, \ldots, x_n) \in P.$$

Since P is r.e., so is R (by (2) of Prop. 1.). We have thus proved Theorem 1.

Since the systems (R), (Q) and P.A. are axiomatizable, then by Theorem 1, any set or relation representable in any of these systems is r.e. Also, by Shepherdson's theorem (or by the Ehrenfeucht-Feferman theorem, whose proof will be given later), all r.e. relations are representable in any consistent axiomatizable extension of Robinson's system (R), hence in (R) itself and in (Q) and in P.A. (assuming P.A. is consistent). Since (Q) is finitely axiomatizable, then there is at least one finitely axiomatizable system in which all r.e. relations are representable. And so we have:

Theorem 2. *For any set or relation R, the following conditions are all equivalent:*

(1) *R is recursively enumerable.*
(2) *R is representable in some finitely axiomatizable system.*
(3) *R is representable in (Q) (or even in (Q_0)).*
(4) *R is representable in (R) (or even in (R_0)).*
(5) *R is representable in P.A. (assuming P.A. consistent).*

We now see that the systems (R), (Q) and P.A., though of different strengths with respect to provability, are all of the same strength with respect to representability (assuming P.A. consistent). The relations representable in any one of these systems are precisely the r.e. relations.

Theorem 3. *If S is consistent and axiomatizable, then every rela- tion (and set) definable in S is recursive.*

Proof. Suppose S is consistent and axiomatizable and that R is defin- able in S. Since S is consistent, then R is completely representable in S (cf. §3, Ch. 0). Hence R and its complement are both repre- sentable in S (by Theorem 1) and so R is recursive.

Corollary. *A relation is recursive iff it is definable in (R). The same is true for the system (Q) and for P.A., assuming P.A. is consistent.*

We now see that the r.e. relations are those representable in (Q); the recursive relations are those definable in (Q). The same is true for the system (R).

Chapter II

Undecidability and Recursive Inseparability

I. Undecidability

§1. Some Preliminary Theorems.   We continue to let S be an arbitrary system, P be the set of Gödel numbers of the provable formulas of S and R be the set of Gödel numbers of the refutable formulas of S.

Theorem 1. *The set $\widetilde{P^*}$ is not representable in S.*

Proof. This is the diagonal argument all over again. If $H(v_1)$ represents $\widetilde{P^*}$ and h is the Gödel number of $H(v_1)$, then $H[\bar{h}]$ is provable in S iff $h \notin P^*$ iff $d(h) \notin P$ iff $H[\bar{h}]$ is not provable in S, which is a contradiction.

Theorem 1.1. *If S is consistent, then P^* is not definable in S.*

Proof. Suppose P^* is definable in S. If S were consistent, then P^* would be completely representable in S (cf. §3.1, Ch. 0). Hence $\widetilde{P^*}$ would be representable in S, contrary to Theorem 1. Therefore, if S is consistent, then P^* is not definable in S.

Theorem 1.2.[1] *If the diagonal function $d(x)$ is strongly definable in S and S is consistent, then P is not definable in S.*

Proof. Suppose $d(x)$ is strongly definable in S. Since $P^* = d^{-1}(P)$, then if P were definable in S, P^* would be definable in S (by Th. 11.2, Ch. 0). Hence S would be inconsistent by Theorem 1.1.

Exercise 1. Show that if S is consistent, then R^* is not definable in S.

---

[1] Tarski, 1953

38

Exercise 2. Show that if S is consistent, then no superset of R^* disjoint from P^* is definable in S, and no superset of P^* disjoint from R^* is definable in S.

Exercise 3. Prove that if S is consistent and if the diagonal function is strongly definable in S, then no superset of P disjoint from R is definable in S. [This is stronger than Theorem 1.2.]

§2.   Undecidable Systems.

A system S is said to be *decidable* (or to admit of a *decision procedure*) if the set P of Gödel numbers of the provable formulas of S is a recursive set. It is *undecidable* if P is not recursive.

This meaning of 'undecidable' should not be confused with the meaning of 'undecidable' when applied to a particular formula (as being undecidable in a given system S). To ask of a particular formula, whether or not it is decidable, is meaningless except with reference to a particular system S. On the other hand, the statement that a given system S is decidable is, so to speak, a mass statement about the entire set of formulas provable in S. The notion of a system S being undecidable is a notion of recursive function theory; the notion of a sentence being undecidable in a given system S is not.

Theorem 2. *If all r.e. sets are representable in S, then the complement $\widetilde{P}$ of P is not r.e.*

Proof. Suppose all r.e. sets are representable in S. Suppose $\widetilde{P}$ were r.e. Then the set $\widetilde{P}^*$ would be r.e. (since $\widetilde{P}^* = d^{-1}(\widetilde{P})$, and $d(x)$ is a recursive function). Hence $\widetilde{P}^*$ would be representable in S, contrary to Theorem 1. [We note that $\widetilde{P}^* = \widetilde{P^*}$.]

Remark. The above proof is by contradiction. A more constructive proof is the following.

To say that $\widetilde{P}$ is not r.e. is to say that for any r.e. set A, $A \neq \widetilde{P}$—or equivalently that for any r.e. set A, there is a number n such that

$$n \in P \leftrightarrow n \in A.$$

Well, suppose A is any r.e. set. Then the set A^* is also r.e. By hypothesis, A^* is representable in S. Let $H(v_1)$ be a formula which represents A^* in S, and let h be the Gödel number of $H(v_1)$. Then $H[\bar{h}]$ is provable in S iff its Gödel number is in A. So, if n is the Gödel number of $H[\bar{h}]$, then

$$n \in P \leftrightarrow n \in A.$$

Therefore, for every r.e. set A, $\widetilde{P} \neq A$. Hence $\widetilde{P}$ is not r.e.

Theorem 3. *If all r.e. sets are representable in S, then S is undecidable.*

Proof. Immediate from Theorem 2.

Corollary. *The systems (R) and (Q) are undecidable, and P.A. (if consistent) is undecidable.*

The following theorem provides an even weaker condition that a system be undecidable.

Theorem 4. *If all recursive sets are representable in S, then S is undecidable.*

Proof. Suppose all recursive sets are representable in S. If S was decidable, then P would be recursive. Hence $\widetilde{P}$ would be recursive, $\widetilde{P^*}$ would be recursive, and $\widetilde{P^*}$ would be representable in S, contrary to Theorem 1. Therefore S is undecidable.

Exercise 4. Prove that if the complement of every r.e. set is representable in S, then S is not axiomatizable.

Exercise 5. Suppose all recursive functions of one argument are strongly definable in S and that S is consistent. Does it necessarily follow that S is undecidable?

The existence of an axiomatizable but undecidable system (e.g. (Q)) yields one proof of a basic result in recursion theory.

Theorem 5. *There exists a recursively enumerable set that is not recursive.*

Proof. For S, the system (Q) and the set P are r.e. (since (Q) is axiomatizable), but P is not recursive (since (Q) is not decidable).

Many other ways of constructing r.e. sets that are non-recursive will be provided in later chapters.

Corollary. *There exists an arithmetic set that is not r.e.*

Proof. Let A be an r.e. set that is not recursive. Since A is r.e., it is certainly arithmetic; hence $\widetilde{A}$ is arithmetic. But $\widetilde{A}$ is not r.e., since A is not recursive.

Theorem 6. *If S is decidable, then every set representable in S is recursive. Stated otherwise, if some non-recursive set is representable in S, then S is undecidable.*

Proof. If A is representable in $\mathcal{S}$, then for some number h,

$$A = x : r(h, x) \in P$$

(cf. proof of Theorem 1). If $\mathcal{S}$ is decidable, then P is recursive; hence A is recursive (by 2 of Proposition 1, Chapter 1).

Discussion. Theorem 6 and Theorem 5 together provide another proof of Theorem 3. Suppose all r.e. sets are representable in $\mathcal{S}$. By Theorem 5, some r.e. set is not recursive. Hence some non-recursive set is representable in $\mathcal{S}$. Then $\mathcal{S}$ is undecidable by Theorem 6.

Of course, this argument does not establish the stronger result (Theorem 4) that if all *recursive* sets are representable in $\mathcal{S}$, then $\mathcal{S}$ is undecidable.

§3. Non-recursivity and Incompleteness.

The existence of a recursively enumerable set which is not recursive affords an interesting alternative approach to Gödel Incompleteness Theorem.

Theorem 7. *If $\mathcal{S}$ is a consistent axiomatizable system in which some non-recursive set is representable, then $\mathcal{S}$ is incomplete.*

Proof. Suppose that $\mathcal{S}$ is a consistent axiomatizable system and that $F(v_1)$ represents A in $\mathcal{S}$ where A is not a recursive set. Let B be the set represented by $\sim F(v_1)$. Since $\mathcal{S}$ is axiomatizable, A and B are both r.e. sets (by Theorem 1). Since A is not recursive, B is not the complement of A. Therefore, some number n is outside both A and B, and neither $F(\bar{n})$ nor $\sim F(\bar{n})$ is provable in $\mathcal{S}$.

Discussion. Theorem 7 provides another proof of the incompleteness of P.A. Assuming P.A. is consistent, all r.e. sets are representable in P.A. Since there exists an r.e. set that is not recursive, some non-recursive set A is representable in P.A. Hence, P.A. is incomplete by Theorem 7 (since P.A. is axiomatizable).

The above incompleteness proof, however, is non-constructive. That is, it shows that there *is* an undecidable sentence of P.A., but it gives no indication how to find one. If we could *find* a number n outside the sets A and B of the proof of Theorem 7, then we could find an undecidable sentence of P.A. (namely $H(\bar{n})$), but we are given no recipe for finding such an n.

This situation will be improved in Chapter 4 in which we will consider an r.e. but non-recursive set C of a very special kind (a so-called *creative* set) having the property that given any formula $H(v_1)$ representing C, we can actually find a number outside both C and the set represented by $\sim H(v_1)$.

The existence of an arithmetic set that is not r.e. provides yet another proof of the incompleteness of P.A. (under the assumption that P.A. is correct). Suppose S is any axiomatizable subsystem of $\mathcal{N}$. Let B be an arithmetic set that is not r.e. Since B is arithmetic, it is expressed by some formula $H(v_1)$. The formula $H(v_1)$ also *represents* some set B_0 in S. Since S is axiomatizable, B_0 is r.e., and since B is not r.e., $B_0 \neq B$. But $B_0 \subseteq B$ because S is correct (it is a subsystem of $\mathcal{N}$); hence $n \in B_0 \Rightarrow H(\bar{n})$ provable in $S \Rightarrow H(\bar{n})$ true $\Rightarrow n \in B$. Therefore, B_0 is a proper subset of B. Hence some number n in B is not in B_0, and for any such number n, the sentence $H(\bar{n})$ is true but not provable in S. This proves:

Theorem 8. *For any axiomatizable subsystem S of $\mathcal{N}$ and for any arithmetic set B that is not r.e. and for any formula $H(v_1)$ expressing B, there is a number n in B such that $H(\bar{n})$ is true but not provable in S.*

Theorem 8, like Theorem 7, has a constructive analogue due to Kleene which will be given in Chapter 5.

§4. Essential Undecidability.
A system S is called *essentially undecidable* if S is consistent, and every consistent extension of S (including S itself) is undecidable.

Theorem 9. *If all recursive sets are definable in S and S is consistent, then S is essentially undecidable (Putnam, 1957).*

Proof. Suppose S is consistent and that every recursive set is definable in S. Let S' be any consistent extension of S. Then all recursive sets are definable in S' (because any formula that defines a set in S also defines it in S'). But since S' is consistent, all recursive sets are then completely representable; hence they are representable in S'. Therefore, S' is undecidable by Theorem 4.

Corollary 1. *Every consistent Rosser system is essentially undecidable.*

Corollary 2. *The system (R) is essentially undecidable. The same is true with the systems (Q) and P.A. (assuming P.A. is consistent).*

Exercise 6. By the characteristic function $C_A(x)$ of a set A, we mean the function that assigns 1 to every element in A and 0 to every element not in A (thus $C_A(x) = 1 \leftrightarrow x \in A$; $C_A(x) = 0 \leftrightarrow x \notin A$).

1. Prove that a set is recursive iff its characteristic function is recursive.
2. Prove that for any set A, if its characteristic function is definable in S, then A is definable in S.
3. Using (a), (b) and Theorem 4, prove the following theorem: If every recursive function of one argument is definable in S, then S is essentially undecidable.

II. *Recursive Inseparability*

§5. **Recursive Inseparability.** The notion of recursive inseparability plays a fundamental role in metamathematics and recursive function theory.

Two disjoint sets A and B are called *recursively separable* if there is a recursive superset A' of A disjoint from B. [This implies that there is a recursive superset of B—namely $\widetilde{A'}$—disjoint from A, so the condition is symmetric]. A and B are called *recursively inseparable* (abbreviated R.I.) iff they are disjoint and not recursively separable.

Exercise 7. For any disjoint sets A and B, show that the following four conditions are all equivalent.

1. Given any r.e. superset A' of A disjoint from B, the complement of A' is not r.e.
2. Given any disjoint r.e. supersets A' and B' of A and B respectively, there is a number outside $A' \cup B'$.
3. Given any r.e. supersets A' and B' of A and B respectively, there is a number n such that $n \in A' \leftrightarrow n \in B'$.
4. A and B are recursively inseparable.

A system S is said to be R.I. or an R.I. system, if the sets P and R are R.I.

Discussion. Here is one reason why recursive inseparability is important in metamathematics (several other reasons will emerge in later chapters). Suppose we have an undecidable system S. Then there is no effective test as to which sentences are provable in S. Now, suppose the pair (P, R) were recursively separable. This means that we could partition the set N of natural numbers into two recursive sets, A and $\tilde{A}$, with P being a subset of A and R being a subset of $\tilde{A}$. Then given a sentence X, we could take its Gödel number n and effectively test whether n belongs to A or whether it belongs to $\tilde{A}$. If we find that n belongs to $\tilde{A}$, then we will at least know that X is *not* provable in S (it is either refutable or undecidable in S), and, hence, we won't waste any time trying to prove it in S. On the other hand, if n turns out to be in A, then we will at least know that X is not refutable (it is either provable or undecidable). Hence we won't lose any time trying to disprove it in S. And so a recursive separation of P from R, though not as good as decidability, would at least give us partial information about provability and refutability in the system. But as we shall shortly see, even this partial information is denied us for Peano Arithmetic.

Theorem 10. *If S is R.I., then S is essentially undecidable.*

Proof. Suppose S is not essentially undecidable. If S is inconsistent, then $P = R$; hence S is not R.I. Suppose S is consistent. Then some consistent extension S' of S is decidable. Let P' and R' be the sets of Gödel numbers of the formulas provable and refutable in S' respectively. Since S' is consistent, P' is disjoint from R'; hence P' is disjoint from R. Since S' is decidable, P' is recursive, so P' is a recursive superset of P disjoint from R, which means P and R are recursively separable. Therefore, if S is R.I., then S is essentially undecidable.

We now wish to show that if all recursive sets are definable in S and S is consistent, then S is not only essentially undecidable (Theorem 9) but is even R.I. This will follow from our next two theorems.

Theorem 11. *No superset of R^* disjoint from P^* is definable in S. Stated otherwise, any definable superset of R^* contains an element of P^*.*

Proof. Suppose $R^* \subseteq A$ and A is defined by $H(v_1)$ in S. Let h be the Gödel number of $H(v_1)$. We show that h is in both A and P^*.

Suppose $h \notin A$. Then $H(\bar{h})$ is refutable (since $H(v_1)$ defines A); hence $H[\bar{h}]$ is refutable and $h \in R^*$, contrary to the fact that $R^* \subseteq A$. Therefore, h must be in A.

Since $h \in A$, $H(\bar{h})$ is provable; hence $h \in P^*$. So h is in both A and P^*.

Theorem 12. (1) *For any recursive function $\phi(x)$ and any disjoint sets A and B, if the sets $\phi^{-1}(A)$ and $\phi^{-1}(B)$ are R.I., then so are the sets A and B.*

(2) *For any system S, if the sets P^* and R^* are R.I., then so are the sets P and R.*

Proof.

1. Let $\phi(x)$ be any recursive function. Suppose A and B are recursively separable. Then there is a recursive superset A' of A disjoint from B and $\phi^{-1}(A')$ is a superset of $\phi^{-1}(A)$ disjoint from $\phi^{-1}(B)$. (Why?) But since A' is recursive, so is $\phi^{-1}(A')$. Thus $\phi^{-1}(A)$ is recursively separable from $\phi^{-1}(B)$. This proves that if A and B are recursively separable, then so are $\phi^{-1}(A)$ and $\phi^{-1}(B)$. Equivalently, if $\phi^{-1}(A)$ and $\phi^{-1}(B)$ are recursively inseparable, then so are A and B.

2. Since the diagonal function $d(x)$ is recursive, $P^* = d^{-1}(P)$ and $R^* = d^{-1}(R)$. The result follows from (a).

Theorem 13. *If all recursive sets are definable in S and S is consistent, then S is R.I. (Smullyan, 1959).*

Proof. Assume hypothesis. Since S is consistent, the pairs (P, R) and (P^*, R^*) are disjoint.

Suppose P^* and R^* were recursively separable. Then some recursive superset A of R^* is disjoint from P^*. By hypothesis, A would be definable in S, contrary to Theorem 11. Therefore, (P^*, R^*) is R.I. Then by (b) of Theorem 12, the pair (P, R) is also R.I.

Theorem 14. *There exists an R.I. pair of r.e. sets.*

Proof. (Q) is a consistent axiomatizable system in which all recursive sets are definable. Therefore, (P, R) and (P^*, R^*) are both examples of an R.I. pair of r.e. sets.

Remark. Several other methods of constructing R.I. pairs of r.e. sets will be provided in Chapter 5.

§6.   Recursive Inseparability and Incompleteness.

Theorem 15. *If some R.I. pair (A, B) of sets is strongly separable in S and S is consistent and axiomatizable, then S is incomplete.*

Proof. Assume hypothesis. Let $H(v_1)$ be a formula that strongly separates A from B in S. Then $H(v_1)$ represents some superset A' of A, and $\sim H(v_1)$ represents some superset B' of B. Since S is consistent, A' is disjoint from B'. Since S is axiomatizable, the sets A' and B' are both r.e., and since the pair (A, B) is recursively inseparable, B' is not the complement of A' (for if it were, A' would be recursive, hence A' would be a recursive superset of A disjoint from B). Therefore there is some n outside both A' and B'. Hence $H(\bar{n})$ is neither provable nor refutable in S.

Discussion. Theorem 15, together with the fact that there exists an R.I. pair of r.e. sets, provides yet another way of proving that any consistent axiomatizable Rosser system for sets (such as Peano Arithmetic) must be incomplete (because in a Rosser system for sets, *every* pair of r.e. sets is strongly separable. Hence some R.I. pair of r.e. sets is strongly separable). However, this proof, like the proof of Theorem 7, is non-constructive since the proof of Theorem 15 doesn't provide any way of *finding* a number outside the sets A' and B'. A constructive version of Theorem 15 (Kleene's Symmetric Form of Gödel's Theorem) will be given in Chapter 5. Indeed, many of the results of this chapter have "effective" analogues, but we cannot state these before turning to the subject of *indexing*, which we will do in the next chapter.

Exercise 8—Incompleteness and Undecidability. Another approach to incompleteness is via undecidability and axiomatizability. Using the facts that the set S of Gödel numbers of sentences is recursive[2] and that for any system S, the set P is recursive iff $P \cap S$ is recursive, fill in the following steps of the proof that if S is axiomatizable but undecidable, then S must be incomplete.

1. Suppose that S is axiomatizable. Then the sets $P \cap S$ and $R \cap S$ are both r.e.
2. Suppose S is also complete. Show that the complement of $P \cap S$ is $R \cup \tilde{S}$. Show that $P \cap S$ is, therefore, recursive and, hence, P must be recursive.
3. Show that it, therefore, follows that if S is axiomatizable but undecidable, then S must be incomplete.

Exercise 9—Church's Theorem. For any system S and any sentence X, by $S + \{X\}$, we mean the system whose axioms are those of S together with X. By a well known result known as the *deduction*

---

[2] Exercise 7, Ch. 4, G.I.T.

theorem, a formula Y is provable in $S + \{X\}$ iff the formula $X \supset Y$ is provable in S.

Now let S_0 be the system whose only axioms are those of first-order logic with identity and function symbols (the arithmetic significance of the function symbols plays no role). One form of the result known as *Church's Theorem* (also anticipated in Gödel, 1931, Proposition IX and Proposition X) is that S_0 is undecidable (there is no decision procedure for first-order logic with identify and function symbols). Using the deduction theorem, this follows from the existence of a finitely axiomatizable but undecidable system (e.g. (Q)) by the following argument (whose steps are to be filled in by the reader).

1. Let X be the conjunction of the closures of the nine non-logical axioms of (Q). Then the provable formulas of (Q) are the provable formulas of $S_0 + \{X\}$.
2. Then by the deduction theorem, a formula Y is provable in (Q) iff $X \supset Y$ is provable in S_0.
3. For any number y, the Gödel number of $X \supset E_y$ is a recursive function of y—call this function $\phi(y)$.
4. Let P_0 be the set of Gödel numbers of the provable formulas of S_0, and let P be the set of Gödel numbers of the provable formulas of (Q). Show then that $P = \phi^{-1}(P_0)$.
5. Show, therefore, that if S_0 were decidable, (Q) would be decidable.

Note. Church's theorem is known in the stronger form that *pure* first-order logic (i.e., first-order logic without identity or function symbols—just predicate symbols) is undecidable. Indeed, pure first-order logic with only one binary predicate symbol is undecidable—for a neat proof, cf. Boolos and Jeffrey, Chapter 22.

Chapter III

Indexing

For the remaining chapters, we will need two basic theorems in recursive function theory—the *enumeration* theorem of Kleene and Post and the *iteration* theorem of Kleene.

I. *The Enumeration Theorem*

§1. Indexing.   We wish to arrange all r.e. sets in an infinite sequence $\omega_0, \omega_1, \ldots, \omega_n, \ldots$ (allowing repetitions) in such a way that the relation $x \in \omega_y$ is r.e.

We shall take the system (Q) as our basic formalism for recursive function theory. We know that (Q) is axiomatizable and that the representable sets of (Q) are precisely the r.e. sets. We define ω_i as the set of all numbers n such that $E_i[\bar{n}]$ is provable in (Q). Equivalently, ω_i is the set of all n such that $r(i,n) \in P$, where $r(i,n)$ is the Gödel number of $E_i[\bar{n}]$ and P is the set of Gödel numbers of the provable formulas of (Q). Since $r(x,y)$ is a recursive function and P is an r.e. set, then the relation $r(x,y) \in P$ is r.e., and this is the relation $y \in \omega_x$. Also, every r.e. set A is represented in (Q) by some formula $E_i(v_1)$; hence $A = \omega_i$. Thus every r.e. set appears in our enumeration.

We call i an *index* of an r.e. set A if $A = \omega_i$. We let $U(x,y)$ be the relation $x \in \omega_y$, and we see that this relation is r.e.

Indexing of r.e. Relations. For each $n \geq 2$, we also wish to arrange all r.e. relations of degree n an in infinite sequence

$$R_0^n, R_1^n, \ldots, R_n^n, \ldots$$

in such a manner that the relation $R_y^n(x_1, \ldots, x_n)$ is an r.e. relation among $x_1, \ldots, x_n$ and y. To this end, it will be convenient to use the indexing of r.e. sets that we already have and to use the recursive pairing function $J(x, y)$ and its associated functions $J_n(x_1, \ldots, x_n)$ (cf. §4, Chapter 1).

We simply define $R_i^n(x_1, \ldots, x_n)$ iff $J_n(x_1, \ldots, x_n) \in \omega_i$. Since ω_i is r.e. and $J_n(x_1, \ldots, x_n)$ is a recursive function, then the relation R_i^n is r.e. (Proposition 2, Chapter 1). Also, given any r.e. relation $R(x_1, \ldots, x_n)$, the set A of all numbers $J_n(x_1, \ldots, x_n)$ such that $R(x_1, \ldots, x_n)$ is r.e. (by (d) of Proposition 2, Chapter 1), and $A = \omega_i$ for some i. Therefore

$$R(x_1, \ldots, x_n) \leftrightarrow J_n(x_1, \ldots, x_n) \in \omega_i \leftrightarrow R_i^n(x_1, \ldots, x_n),$$

and so $R = R_i^n$. Thus, every r.e. relation has an index. Finally, since the relation $x \in \omega_y$ is r.e. and $J_n(x_1, \ldots, x_n)$ is a recursive function, the relation $J_n(x_1, \ldots, x_n) \in \omega_y$, as a relation among $x_1, \ldots, x_n$ and y, is an r.e. relation, and this is the relation $R_y^n(x_1, \ldots, x_n)$.

Remark. An alternative scheme for indexing r.e. relations of degree ≥ 2 and one that does not use the functions $J_n(x_1, \ldots, x_n)$ is this: For each $n \geq 2$, we use the function $r_n(x, y_1, \ldots, y_n)$ of §6 of Chapter 1 ($r_n(x, y_1, \ldots, y_n)$ is the Gödel number of $E_x[\bar{y}_1, \ldots, \bar{y}_n]$). We define R_i^n as the set of all n-tuples $(x_1, \ldots, x_n)$ such that

$$r_n(i, x_1, \ldots, x_n) \in P.$$

However, the scheme we have chosen has the technically handy property that for each $n \geq 2$,

$$R_i^n(x_1, \ldots, x_n) \leftrightarrow J_n(x_1, \ldots, x_n) \in \omega_i.$$

We have now proved:

Theorem 1—The Enumeration Theorem.   *For each $n \geq 1$, there exists an enumeration $R_0^n, R_1^n, \ldots, R_i^n, \ldots$ of all r.e. relations of degree n such that the set of all $(n+1)$-tuples $(x_1, \ldots x_n, y)$, for which $R_y^n(x_1, \ldots, x_n)$ holds, is an r.e. relation.*

For each $n \geq 1$, we let $U^{n+1}(x_1, \ldots, x_n, y)$ be the relation $R_y^n(x_1, \ldots, x_n)$. These relations are sometimes referred to as the *universal* relations ($U^2(x, y)$ is the universal relation for r.e. sets, $U^3(x_1, x_2, y)$ is the universal relation for r.e. relations of degree 2, and so forth). No ambiguity will result if we write $R_i(x_1, \ldots, x_n)$ for $R_i^n(x_1, \ldots, x_n)$ and $U(x_1, \ldots, x_n, y)$ for $U^n(x_1, \ldots, x_n, y)$, since the number of arguments makes the superscript clear. Also "ω_i" is synonymous with "R_i^1".

Exercise 1.

1. Given a collection C of sets of natural numbers, a binary relation $R(x, y)$ on the natural numbers is said to *enumerate* C if:

   a. For every set A in C there is a number n such that for all x, $x \in A \leftrightarrow R(x, n)$;
   b. For each n, $\{x : R(x, n)\}$ is a member of C.

   By the enumeration theorem, there is an r.e. relation $U(x, y)$ that enumerates all r.e. sets. Show that there is no arithmetic relation $A(x, y)$ that enumerates all arithmetic sets.
2. Let C be the set of all numbers x such that $x \in \omega_x$. [This set C will play a key role in the next chapter.] The set C is obviously r.e. Is it recursive?
3. Let C be a collection of sets and relations of natural numbers such that for any relation $R(x, y)$, if $R \in C$, then so is

$$x : R(x, x).$$

   Prove that the following two conditions are logically incompatible:

   a. There is a relation $R(x, y)$ in C that enumerates all the sets in C;
   b. the complement of every set in C is again in C.

4. Show that the answers to 1 and 2 both follow from 3.

Exercise 2. Is it possible to enumerate all recursive functions of one argument in a sequence $f_0(x), f_1(x), \ldots, f_n(x), \ldots$ in such a way that the function $f_x(y)$ (as a function of the two variables x and y) is a recursive function?

Exercise 3. Show that there is a recursive function $\phi(x)$ with the following two properties:

1. For any number i, $\phi(i)$ is the Gödel number of a formula that represents ω_i in (Q);
2. For every i, $\phi(i) > i$.

Exercise 4. Show that every r.e. set has infinitely many indices.

II.   The Iteration Theorem

§2.   The Iteration Theorem.

A second basic tool of recursive function theory is the iteration theorem (sometimes called the S_n^m theorem) due to Kleene (also discovered independently by Martin Davis). To fully appreciate the power and significance of this theorem, the reader should first try the following exercises.

Exercise 5. We know that for any two r.e. sets ω_i and ω_j, their union $\omega_i \cup \omega_j$ is r.e. Is there a recursive function $\phi(x, y)$ such that for any numbers i and j,

$$\omega_{\phi(i,j)} = \omega_i \cup \omega_j?$$

[The question can be paraphrased: Given two r.e. sets (in the sense that we are given indices for them) can we effectively find an index of their union?]

Exercise 6. Show that there is a recursive function $\phi(x)$ such that for any numbers i, x and y

$$R_{\phi(i)}(x, y) \leftrightarrow R_i(y, x).$$

[This can be paraphrased: Given an index of an r.e. relation of two arguments, we can "find" an index of its inverse.]

Exercise 7. Show that for any recursive function $f(x)$ there is a recursive function $t(x)$ such that for every number i,

$$\omega_{t(i)} = f^{-1}(\omega_i).$$

The reader who has solved the above problems will have found himself repeatedly going back to the formal system (Q). The iteration theorem (Theorem 2 below) frees us from this once and for all.

Theorem 2—Part 1. *For any r.e. relation $R(x, y)$, there is a recursive function $\phi(y)$ such that for all numbers i and x,*

$$x \in \omega_{\phi(i)} \leftrightarrow R(x, i).$$

Proof. Suppose $R(x, y)$ is r.e. Then it is represented in (Q) by some formula $F(v_1, v_2)$. For any number n, let $F[v_1, \bar{n}]$ be the formula $\forall v_2[v_2 = \bar{n} \supset F(v_1, v_2)]$ and let $\phi(n)$ be its Gödel number. The function ϕ is easily seen to be recursive. For any number i, the formula $F[v_1, \bar{i}]$ represents $x : R(x, i)$ and so its Gödel number $\phi(i)$ is an index of this set. Thus $x \in \omega_{\phi(i)} \leftrightarrow R(x, i)$.

Theorem 2—Part 2. *For any positive integers m and n, and any r.e. relation $R(x_1, \ldots, x_m, y_1, \ldots, y_n)$, there is a recursive function $\phi(y_1, \ldots, y_n)$ such that for all numbers $x_1, \ldots, x_m$ and $i_1, \ldots, i_n$:*

$$R_{\phi(i_1, \ldots, i_n)}(x_1, \ldots, x_m) \leftrightarrow R(x_1, \ldots, x_m, i_1, \ldots, i_n).$$

Proof. This can be obtained from Part I and Proposition 10 of Chapter 1 as follows: By Proposition 10, Chapter 1, there is an r.e. relation $M(x, y)$ such that for all numbers $x_1, \ldots, x_m$ and $y_1, \ldots, y_n$,

(1) $R(x_1, \ldots, x_m, y_1, \ldots, y_n) \leftrightarrow M(J_m(x_1, \ldots, x_m), J_n(y_1, \ldots, y_n)).$

Then by Part 1 applied to $M(x, y)$, there is a recursive function $t(y)$ such that for all x and y, $x \in \omega_{t(y)} \leftrightarrow M(x, y)$. Therefore, for all $x_1, \ldots, x_m$ and $y_1, \ldots, y_n$ we have:

(2) $J_m(x_1, \ldots, x_m) \in w_{tJ(y_1, \ldots, y_n)}$
$\qquad\qquad \leftrightarrow M(J_m(x_1, \ldots, x_m), J_n(y_1, \ldots, y_n))$
$\qquad\qquad \leftrightarrow R(x_1, \ldots, x_m, y_1, \ldots, y_n).$

Hence

$R_{tJ(y_1, \ldots, y_n)}(x_1, \ldots, x_m) \quad \leftrightarrow \quad J_m(x_1, \ldots, x_m) \in \omega_{tJ(y_1, \ldots, y_n)}$
$\qquad\qquad\qquad\qquad\qquad \leftrightarrow \quad R(x_1, \ldots, x_m, y_1, \ldots, y_n).$

And so we take $\phi(y_1, \ldots, y_n) = tJ(y_1, \ldots, y_n).$

Remark. If we had taken the alternative indexing scheme mentioned in §1 (the one that doesn't employ a pairing function), then we would have proved Part 2 by taking a formula

$$F(v_1, \ldots, v_m, v_{m+1}, \ldots, v_n)$$

that represents

$$R(x_1, \ldots, x_m, y_1, \ldots, y_n)$$

in (Q) and by taking $\phi(y_1, \ldots, y_n)$ to be the Gödel number of

$$F[v_1, \ldots, v_n, \overline{y}_1, \ldots, \overline{y}_n].$$

Informally, the iteration theorem says that given an r.e. relation $R(x_1, \ldots, x_m, y_1, \ldots, y_n)$, if we plug in numbers $i_1, \ldots, i_n$ for the variables $y_1, \ldots, y_n$, we can find an index of the relation

$$R(x_1, \ldots, x_m, i_1, \ldots, i_n)$$

as a recursive function of $i_1, \ldots, i_n$.

Applications of the Iteration Theorem. Using the iteration theorem, let us now solve the last three exercises.

For Exercise 5, let $R(x, y_1, y_2)$ be the r.e. relation:

$$x \in \omega_{y_1} \lor x \in \omega_{y_2}.$$

Then by the iteration theorem, there is a recursive function $\phi(y_1, y_2)$ such that for all numbers i_1, i_2, x:

$$x \in \omega_{\phi(i_1, i_2)} \leftrightarrow R(x, i_1, i_2) \leftrightarrow (x \in \omega_{i_1} \lor x \in \omega_{i_2}).$$

Therefore $\omega_{\phi(i_1, i_2)} = \omega_{i_1} \cup \omega_{i_2}$.

For Exercise 6, let $R(x_1, x_2, y)$ be the r.e. relation $R_y(x_2, x_1)$. [This relation is explicitly definable from the relation $U^3(x_1, x_2, y)$. Hence it is r.e.]. Then by the iteration theorem, there is a recursive function $\phi(y)$ such that for all x_1, x_2 and i,

$$R_{\phi(i)}(x_1, x_2) \leftrightarrow R(x_1, x_2, i) \leftrightarrow R_i(x_2, x_1).$$

For Exercise 7, given a recursive function $f(x)$, let $R(x, y)$ be the r.e. relation $f(x) \in \omega_y$. Then by the iteration theorem, there is a recursive function $\phi(y)$ such that for all x and i,

$$x \in \omega_{\phi(i)} \leftrightarrow R(x, i) \leftrightarrow f(x) \in \omega_i.$$

Hence, $\omega_{\phi(i)} = f^{-1}(\omega_i)$.

Other applications of the iteration theorem are given in some of the exercises that follow.

Exercise 8. 1. Prove that there is a recursive function $\phi(y)$ such that for all i and x, $x \in \omega_{\phi(i)} \leftrightarrow R_i(x, x)$.
2. Prove that there is a recursive function $\phi(y)$ such that for all i and x, $x \in \omega_{\phi(i)} \leftrightarrow \exists y R_i(x, y)$.
3. Prove that there is a recursive function $\phi(y_1, y_2)$ such that for any recursive functions $f(x)$ and $g(x)$ and any numbers i and j, if i is an index of the relation $f(x) = y$ and j is an index of the relation $g(x) = y$, then $\phi(i, j)$ is an index of the relation $f(g(x)) = y$.
4. For any two sets A and B, their *cross-product* $A \times B$ is defined to be the set of all numbers $J(x, y)$ such that $x \in A$ and $y \in B$.
   Prove that there is a recursive function $\phi(y_1, y_2)$ such that for all numbers i_1 and i_2, $\omega_{\phi(i_1, i_2)} = \omega_{i_1} \times \omega_{i_2}$.

Exercise 9. Show that for any positive m and n and any r.e. relation $R(x_1, \ldots, x_m, y_1, \ldots, y_n)$ there is a recursive function

$$\phi(x_1, \ldots, x_m)$$

such that for all numbers $i_1, \ldots, i_m$ and $y_1, \ldots, y_n$:

$$R_{\phi(i_1,\ldots,i_m)}(y_1, \ldots, y_n) \leftrightarrow R(i_1, \ldots, i_m, y_1, \ldots, y_n).$$

Exercise 10. A denumerable sequence $A_1, A_2, \ldots, A_m, \ldots$ consisting of all r.e. sets is called an *r.e. indexing* if the relation $x \in A_y$ is r.e. It is called a *maximal* indexing if for every r.e. indexing $B_1, B_2, \ldots, B_n, \ldots$, there is a recursive function $\phi(x)$ such that for all $i, A_{\phi(i)} = B_i$. [Informally, this says that given any B-index of an r.e. set, we can effectively find an A-index of that set.]

By the enumeration theorem, the sequence $\omega_1, \omega_2, \ldots, \omega_n, \ldots$ is an r.e. indexing. Using the iteration theorem, show that it is maximal. [Incidentally, it is known that not every r.e. indexing is maximal.]

Exercise 11—The Uniform Iteration Theorem. Show that for any positive m and n, there is a recursive function $\phi(y, y_1, \ldots, y_n)$ such that for all numbers $i, i_1, \ldots, i_n$, and $x_1, \ldots, x_m$,

$$R_{\phi(i,i_1,\ldots,i_n)}(x_1, \ldots, x_m) \leftrightarrow R_i(x_1, \ldots, x_m, i_1, \ldots, i_n).$$

[This can be derived as a corollary of the iteration theorem together with the enumeration theorem.]

Exercise 12. For any number n (by $\{n\}$, we mean the set whose only member is n), show that there is a recursive function $\phi(x)$ such that for any number n, $\omega_{\phi(n)} = \{n\}$.

More generally, show that for any recursive function $f(x)$, there is a recursive function $\phi(x)$ such that for any n, $\omega_{\phi(n)} = \{f(n)\}$.

Exercise 13. Prove that there is a recursive function $\phi(y_1, y_2)$ such that for any numbers i and j, $\omega_{\phi(i,j)} = \omega_i \cup \{j\}$.

Exercise 14. Prove that for any r.e. set A, there is a recursive function $\phi(x)$ such that for any number i in A, $\omega_{\phi(i)}$ is the set of all natural numbers, and for any i outside A, $\omega_{\phi(i)}$ is the empty set.

Exercise 15. Prove that for any r.e. set A and any r.e. relation $R(x, y)$, there is a recursive function $\phi(y)$ such that for all i :

1. $i \in A \Rightarrow \omega_{\phi(i)} = x : R(x, i)$.
2. $i \notin A \Rightarrow \omega_{\phi(i)} = \emptyset$. [$\emptyset$ is the empty set.]

Exercise 16. By definition, a set A is r.e. iff it is the existential quantification of a Σ_0-relation $R(x, y)$. Show that given any i, we can effectively find a number j such that $R_j(x, y)$ is a Σ_0-relation whose existential quantification is ω_i. More precisely, prove that there is a recursive function $\phi(y)$ such that for every number i, $R^2_{\phi(i)}$

is Σ_0 and $\omega_i = x : \exists y R_{\phi(i)}(x, y)$.

Master Functions. We shall call a function $F(y, z)$ a *master function* if it is recursive and if for every r.e. relation $M(x, y, z)$, there is a recursive function $t(y)$ such that for all numbers y and z,

$$\omega_{F(t(y), z)} = x \colon M(x, y, z).$$

The following theorem will prove quite useful in later chapters.

Theorem 2.1. *There exists a master function* $F(y, z)$.

Proof. The set of all triples (x, y, z) such that $R_y(x, z)$ is an r.e. relation and, hence, by the iteration theorem, there is a recursive function $F(y, z)$ such that for all x, y and z,

(1) $$x \in \omega_{F(y, z)} \leftrightarrow R_y(x, z).$$

We show that F is a master function. Let $M(x, y, z)$ be an r.e. relation. Take the set of all triples (y, x, z) such that $M(x, y, z)$—this set is an r.e. relation. By the iteration theorem, there is a recursive function $t(y)$ such that for all x, y and z,

(2) $$R_{t(y)}(x, z) \leftrightarrow M(x, y, z).$$

In (1) we replace y by $t(y)$, and we have

(1)′ $$x \in \omega_{F(t(y), z)} \leftrightarrow R_{t(y)}(x, z).$$

By (1)′ and (2), it follows that $x \in \omega_{F(t(y), z)} \leftrightarrow M(x, y, z)$, and so $\omega_{F(t(y), z)} = x \colon M(x, y, z)$.

More generally, let us call a function $F(y, z_1, \ldots, z_n)$ a master function if it is recursive, and if for any r.e. relation $M(x, y, z_1, \ldots, z_n)$, there is a recursive function $t(y)$ such that for all $y, z_1, \ldots, z_n$:

$$\omega_{F(t(y), z_1, \ldots, z_n)} = x \colon M(x, y, z_1, \ldots, z_n).$$

If, in the proof of Theorem 2.1, we replace "z" by "$z_1, \ldots, z_n$" everywhere, we will have a proof of:

Theorem 2.2. *For each positive* n, *there exists a master function* $F(x, y, z_1, \ldots, z_n)$ *of* $n + 2$ *arguments.*

Exercise 18. Suppose $F(y, z_1, \ldots, z_n)$ is a master function. Prove that for every r.e. relation $M(x, y_1, \ldots, y_k, z_1, \ldots, z_n)$, there is recursive function $t(y_1, \ldots, y_k)$ such that for all $y_1, \ldots, y_k$ and $z_1, \ldots, z_n$:

$$\omega_{F(t(y_1,\ldots,y_k),z_1,\ldots,z_n)} = x \colon M(x, y_1, \ldots, y_k, z_1, \ldots, z_n).$$

Exercise 19. Prove the more general fact that if $F(y, z_1, \ldots, z_n)$ is a master function, then for any positive integers k and r and any r.e. relation $M(x_1, \ldots, x_r, y_1, \ldots, y_k, z_1, \ldots, z_n)$, there is a recursive function $t(y_1, \ldots, y_k)$ such that for all $x_1, \ldots, x_r, y_1, \ldots, y_k$ and $z_1, \ldots, z_n$,

$$R_{F(t(y_1,\ldots,y_k),z_1,\ldots,z_n)}(x_1, \ldots, x_r) \;\leftrightarrow$$
$$M(x_1, \ldots, x_r, y_1, \ldots, y_k, z_1, \ldots, z_n).$$

III.   Effective Separation

The results that follow will have many applications.

§3.   Uniform Separation.   The next theorem is a "uniform" version of Theorem 10, Chapter 0.

Theorem 3. *There is an r.e. relation $B(x, y, z)$ (which we read "$x \in \omega_y$ before $x \in \omega_z$") such that for all numbers i and j,*

1. $x : B(x, i, j)$ *is disjoint from* $x : B(x, j, i)$,
2. $\omega_i - \omega_j \subseteq x : B(x, i, j)$ *and* $\omega_j - \omega_i \subseteq x : B(x, j, i)$.
3. *If ω_i and ω_j are disjoint, then $\omega_i = x : B(x, i, j)$ and* $\omega_j = x : B(x, j, i)$.

Proof. Since the relation $x \in \omega_y$ is r.e., then there is a recursive relation (in fact a Σ_0-relation) $P(x, y, z)$ such that for all x and y, $x \in \omega_y \leftrightarrow \exists z P(x, y, z)$. [We read $P(x, y, z)$ as "z puts x in ω_y". Then $x \in \omega_y$ iff some z puts x in ω_y.] We now define $B(x, y, z)$ as

$$B(x, y, z) \leftrightarrow \exists w(P(x, y, w) \wedge (\forall w' \leq w) \sim P(x, z, w')).$$

[We can informally read $B(x, y, z)$ as "some number puts x in ω_y and no number less than or equal to it puts x in ω_z"—or, more briefly, "$x \in \omega_y$ before $x \in \omega_z$".] Since $P(x, y, z)$ is recursive, $P(x, y, z)$ and $\sim P(x, y, z)$ are both r.e. Hence $B(x, y, z)$ is easily seen to be r.e. The verifications of (1), (2) and (3) are obvious.

Exercise 20. For each $n \geq 2$, let $B_n(x_1, \ldots, x_n, y, z)$ be the r.e. relation $B(J(x_1, \ldots, x_n), y, z)$. [We read it: "$(x_1, \ldots, x_n) \in R_y^n$ before

$(x_1, \ldots, x_n) \in R_z^n.$] Show that

1. For each i and j, the relations

$$B_n(x_1, \ldots, x_n, i, j) \text{ and } B_n(x_1, \ldots, x_n, j, i)$$

(as relations among $x_1, \ldots, x_n$) are disjoint,
2. $R_i^n - R_j^n \subseteq \lambda x_1 \ldots x_n : B_n(x_1, \ldots, x_n, i, j)$,
3. If R_i^n and R_j^n are disjoint, then

$$R_i^n = \lambda x_1 \ldots x_n : B_n(x_1, \ldots, x_n, i, j).$$

Effective Separation. From Theorem 3 and the iteration theorem, we have:

Theorem 4. *There is a recursive function $\sigma(x, y)$ such that for all i and j,*

1. $\omega_{\sigma(i,j)}$ *and* $\omega_{\sigma(j,i)}$ *are disjoint supersets of* $\omega_i - \omega_j$ *and* $\omega_j - \omega_i$, *respectively.*
2. *If* ω_i *and* ω_j *are disjoint, then* $\omega_i = \omega_{\sigma(i,j)}$ *and* $\omega_j = \omega_{\sigma(j,i)}$.

Proof. We take an r.e. relation $B(x, y, z)$ satisfying Theorem 3. By the iteration theorem, there is a recursive function $\sigma(y_1, y_2)$ such that for all i and j, $\omega_{\sigma(i,j)} = x : B(x, i, j)$. The result then follows by Theorem 3.

We take the function $\sigma(x, y)$ constructed above as fixed throughout this volume.

Exercise 21. Show that for each $n \geq 1$ and every i and j,

1. $R_{\sigma(i,j)}^n$ is disjoint from $R_{\sigma(j,i)}^n$,
2. $R_i^n - R_j^n \subseteq R_{\sigma(i,j)}^n$,
3. If R_i^n and R_j^n are disjoint, then $R_i^n = R_{\sigma(i,j)}^n$.

Chapter IV

Generative Sets and Creative Systems

We now have the background to study the beautiful subject of creative and productive sets inaugurated by Emil Post [1944]. This plays a key rôle in the metamathematical study of incompleteness and undecidability.

§1. Productive and Creative Sets.

To say that a set A is not r.e. is equivalent to saying that for any r.e. subset ω_i of A, there is a number in A not in ω_i. A is called *productive* if there is a recursive function $\phi(x)$—called a *productive function* for A—such that for any number i, if $\omega_i \subseteq A$, then $\phi(i) \in A - \omega_i$. [Informally, this means that it is not only true that no r.e. subset of A is A, but given any such r.e. subset, we can effectively find a number which is in A but not in the subset.] A set A is called *creative* (after Post) if it is r.e., and its complement is productive. Let us note that a productive function for the complement of a set A is a recursive function $\phi(x)$ such that for any number i such that ω_i is disjoint from A, the number $\phi(i)$ lies outside both ω_i and A.

A system $\mathcal{S}$ is called productive if the set P of Gödel numbers of the provable formulas of $\mathcal{S}$ is a productive set; $\mathcal{S}$ is called *creative* if the set P is creative. As we will see, the complete theory $\mathcal{N}$ is not only not axiomatizable but is productive, and the system P.A. is not only undecidable but creative.

Post's Sets C and K. A simple example of a creative set is Post's set C—the set of all numbers x such that $x \in \omega_x$. Thus for any number i, $i \in C \leftrightarrow i \in \omega_i$. If ω_i is disjoint from C, then i is outside both ω_i and C and, therefore, the identity function $I(x)$ is a productive function for $\widetilde{C}$. Therefore, $\widetilde{C}$ is productive, and since C

is obviously r.e., C is creative.

A less immediate example of a creative set is the set K of all numbers $J(x, y)$ such that $y \in \omega_x$. [J is the recursive pairing function.] This set K was also introduced by Post (1943) and is called the *complete* set. The proof that K is creative is less immediate than the proof for C (it involves use of the iteration theorem, as we will see).

Complete Productivity and Creativity. It is also correct to say that a set A is not r.e. iff for every r.e. set ω_i, whether a subset of A or not, there is a number j such that $j \in A \leftrightarrow j \notin \omega_i$. By a *completely productive* function for A, we mean a recursive function $\phi(x)$ such that for every number i, $\phi(i) \in A \leftrightarrow \phi(i) \notin \omega_i$. A set A is called *completely productive* if there is a completely productive function for A. If A is r.e. and $\tilde{A}$ is completely productive, then A is called *completely creative*, and any completely productive function for $\tilde{A}$ is called a completely creative function for A (provided A is r.e.).

The set C of all x such that $x \in \omega_x$ is obviously not only creative but completely creative since the identity function is clearly a completely creative function for C. Therefore, a completely creative set exists. Post apparently did not capitalize on the fact that the set C is not only creative, but completely creative; we hazard a guess that if he had, many results in recursive function theory would have come to light sooner.

Any completely productive function for A is obviously also a productive function for A. By a famous result of John Myhill (which we will study in Chapter 10), if there exists a productive function for A, then there exists a completely productive function for A and, hence, any productive set is completely productive.

Generative Sets. A set whose complement is productive is sometimes called *co-productive*. Sets whose complements are *completely* productive are of sufficient importance to warrant giving them a name; we shall call such sets *generative* sets. (This terminology follows Smullyan [1963].) And by a generative function for A, we mean a completely productive function for $\tilde{A}$. Thus, a generative function for A is a recursive function $\phi(x)$ such that for any number i, $\phi(i) \in A \leftrightarrow \phi(i) \in \omega_i$.

Generative sets will be the main object of study in this chapter. We call a system $\mathcal{S}$ generative if the set P is generative, and we call $\mathcal{S}$ completely creative if $\mathcal{S}$ is generative and axiomatizable—i.e., if P is completely creative.

§2. Many-One Reducibility; Universal Sets. By a

(many-one) *reduction* of a set A to a set B, we mean a recursive function $f(x)$ such that $A = f^{-1}(B)$ (i.e., for all x, $x \in A \leftrightarrow f(x) \in B$).
A set A is said to be (many-one) reducible to B if there is a reduction
of A to B. A function is also said to *reduce* A to B if it is a reduction
of A to B. A set α is called *universal* (sometimes *many-one complete*) if every r.e. set is reducible to α. [Throughout this chapter we
use the term *reducible* to mean many-one reducible. There are also
other important types of reducibility that occur in recursive function
theory.]

Exercise 1. Show that Post's set K is universal.

Proposition 1. *If A is reducible to B and A is generative, then B
is generative.*

It will be useful to state and prove Proposition 1 in a more specific
form. For any r.e. relation $R(x,y)$, let us call a function $t(y)$ an
iterative function for $R(x,y)$ if $t(y)$ is recursive and if for all i,

$$\omega_{t(i)} = x : R(x,i).$$

[The existence of such a function $t(y)$ is guaranteed by the iteration
theorem.] So we will more specifically prove:

Proposition 1*. *If $f(x)$ reduces A to B and $\phi(x)$ is a generative
function for A and $t(y)$ is an iteration function for the relation
$f(x) \in \omega_y$, then the function $f\phi t(x)$ is a generative function for
B.*

Proof. Assume hypothesis. Then for any numbers i and x,

$$x \in \omega_{t(i)} \leftrightarrow f(x) \in \omega_i.$$

Hence for any number i:

$$
\begin{aligned}
f\phi t(i) \in B \;\; &\leftrightarrow \;\; \phi t(i) \in A \quad \text{(since } f \text{ reduces } A \text{ to } B\text{)},\\
&\leftrightarrow \;\; \phi t(i) \in \omega_{t(i)} \quad \text{(since } \phi \text{ is generative for } A\text{)},\\
&\leftrightarrow \;\; f\phi t(i) \in \omega_i \quad \text{(since } t \text{ is iterative for the}\\
&\qquad\qquad\qquad\qquad \text{relation } f(x) \in \omega_y \text{)}.
\end{aligned}
$$

Exercise 2. Show that if A is reducible to B and A is productive,
then B is productive.

Theorem 1. *Every universal set is generative.*

Proof. Suppose α is universal. Post's set C is generative and r.e. Since C is r.e., it is reducible to α (because α is universal). Therefore, α is generative by Proposition 1.

Corollary. *Post's complete set K is generative.*

Proof. By Exercise 1, the set K is universal. [The solution to Exercise 1 is that for any number i, the function $J(i,x)$ (as a function of x) is a reduction of ω_i to K.] Therefore, K is generative by Theorem 1.

Remark. The above proof that K is generative appeals to a previously constructed r.e. generative set—viz. the set C. Another proof will be given shortly.

Exercise 2. Suppose that $f(x)$ is a reduction of Post's creative set C to α and that $t(y)$ is an iterative function for the relation

$$f(x) \in \omega_y.$$

Show that $ft(x)$ is a generative function for α.

Uniform Universality. For any recursive function $f(x,y)$ and any set A, we will say that A is *uniformly universal* under $f(x,y)$ if for every number i, the function $f(i,x)$ (as a function of x) is a reduction of ω_i to A (which means that for all i and x we have the equivalence $x \in \omega_i \leftrightarrow f(i,x) \in A$). We will call A *uniformly universal* if it is universal under some recursive function $f(x,y)$.

Every uniformly universal set is obviously universal (and we will soon see that the converse also holds). The complete set K of Post is not only universal but uniformly universal under the function $J(x,y)$. Taking advantage of the fact that K is *uniformly* universal, we can prove that K is generative without appeal to the existence of a completely creative set—the argument is a special case of the following proposition.

Proposition 2. *If A is uniformly universal under $f(x,y)$ and $t(y)$ is an iterative function for the relation $f(x,x) \in \omega_y$, then $f(t(x),t(x))$ is a generative function for A.*

Proof. Assume hypothesis. Then for any numbers i and x,

1. $f(t(i),x) \in A \leftrightarrow x \in \omega_{t(i)} \leftrightarrow f(x,x) \in \omega_i$.
   Taking $t(i)$ for x we have

2. $f(t(i),t(i)) \in A \leftrightarrow f(t(i),t(i)) \in \omega_i$.

Remark. The above proposition shows that if $t(y)$ is any iterative function for the relation $J(x,x) \in \omega_y$, then $J(t(x), t(x))$ is a generative function for the set K. On the other hand, using our first proof of the generativity of K, we see that if c is any index of the set C and if $t(y)$ is an iterative function for the relation $J(c,x) \in \omega_y$, then $J(c, t(x))$ is a generative function for K. Thus the two proofs yield different generative functions for K.

Proposition 3. *Every universal set is uniformly universal.*

Proof. The set K is uniformly universal under $J(x,y)$. Now, suppose A is universal. Since K is r.e., K is reducible to A under some recursive function $f(x)$. Then A must be uniformly universal under the function $fJ(x,y)$ because for all i and x,

$$x \in \omega_i \leftrightarrow J(i,x) \in K \leftrightarrow fJ(i,x) \in A.$$

§3. Representability and Uniform Representability.

We let S be a system and P the set of Gödel numbers of its provable formulas. We continue to use the recursive function where we recall that $r(x,y)$ ($r(x,y) = $ Gödel number of $E_x[\bar{y}]$).

Proposition 4. *If A is representable in S, then A is reducible to P.*

Proof. If h is the Gödel number of a formula that represents A in S, then the function $r(h,x)$ (as a function of x) is a reduction of A to P because for any number n, $n \in A \leftrightarrow E_h[\bar{n}]$ is provable in $S \leftrightarrow r(h,n) \in P$.

Corollary 1. *If some generative set is representable in S, then S is generative (i.e., P is a generative set).*

Proof. By Proposition 4 and Proposition 1.

Corollary 2. *If all r.e. sets are representable in S, then S is generative.*

Proof. By Corollary 2 and the fact that there is the r.e. generative set C.

We know that all r.e. sets are representable in the systems $(R), (Q)$ and P.A. (if P.A. is consistent) and that these three systems are axiomatizable. Therefore, by Corollary 2, we have:

Theorem 2. *The systems* $(R), (Q)$, *and P.A. (if consistent) are all completely creative. In fact, every consistent axiomatizable extension of* (R) *is completely creative.*

Remark. The above proof of Theorem 2 makes appeal to the existence of a completely creative set. As pointed out by Bernays [1957] in his review of Myhill [1955], a more direct proof is possible and, in fact, yields an alternative proof of the existence of a completely creative set. We shall now turn to this.

Uniform Representability. We shall say that all r.e. sets are *uniformly* representable in S if there is a recursive function $g(x)$ such that for every number i, $g(i)$ is the Gödel number of a formula that represents ω_i in S. [Informally, this means that "given" any r.e. set, we can "find" a formula that represents it.]

Proposition 5. *If the relation* $x \in \omega_y$ *is representable in* S, *then all r.e. sets are uniformly representable in* S.

Proof. Let $\psi(v_1, v_2)$ be a formula that represents the relation $x \in \omega_y$ in S. For any n, we let $\psi[v_1, \bar{n}]$ be the formula

$$\forall v_2(v_2 = \bar{n} \supset \psi(v_1, v_2))$$

and we let $g(n)$ be its Gödel number. The function $g(x)$ is recursive. For any number $n, E_{g(n)}$ is the formula

$$\forall v_2(v_2 = \bar{n} \supset \psi(v_1, v_2)).$$

Hence for any number $m, E_{g(n)}(\bar{m})$ is the sentence

$$\forall v_2(v_2 = \bar{n} \supset \psi(\bar{m}, v_2))$$

and this sentence is provably equivalent to the sentence $\psi(\bar{m}, \bar{n})$. Therefore, for any numbers n and m, $E_{g(n)}(\bar{m})$ is provable in S if, and only if, $\psi(\bar{m}, \bar{n})$ is provable in S if, and only if, $m \in \omega_n$. Therefore, $E_{g(n)}(v_1)$ represents the set ω_n in S.

Corollary. *If* S *is any consistent axiomatizable extension of* (R), *then all r.e. sets are uniformly representable in* S. *In particular, all r.e. sets are uniformly representable in* $(R), (Q)$ *and in P.A. (assuming P.A. is consistent).*

Systems in which all r.e. sets are uniformly representable enjoy the following property.

Theorem 3. *If all r.e. sets are uniformly representable in* S, *then there is a recursive function* $\sigma(x)$ *with the following two properties:*

(1) $\sigma(x)$ *is a generative function for* P.

(2) *For every number* i, $\sigma(i)$ *is the Gödel number of a sentence.*

Proof. Suppose $g(x)$ is a recursive function such that for any number i, $E_{g(i)}(v_1)$ represents ω_i in S. Then for all numbers i and y, $r(g(i), y) \in P \leftrightarrow y \in \omega_i$. Therefore, P is uniformly universal under the function $r(g(x), y)$. Now let $t(y)$ be an iterative function for the relation $r(g(x), x) \in \omega_y$ (thus $\omega_{t(i)} = x : r(g(x), x) \in \omega_i$). Then by Proposition 2, the function $r(gt(x), t(x))$ is a generative function for P. Also, for any number z, $g(z)$ is the Gödel number of a formula in which v_1 is the only free variable. Hence, for any number x, $gt(x)$ is the Gödel number of a formula in the free variable v_1, and $r(gt(x), t(x))$ is the Gödel number of a sentence, so we take $\sigma(x)$ to be $r(gt(x), t(x))$.

We might call a system S *sententially generative* if there is a recursive function $\sigma(x)$ satisfying the conclusion of the above theorem. This means that there is a recursive function $\sigma(x)$ such that for every number i, $E_{\sigma(i)}$ is a sentence and $E_{\sigma(i)}$ is provable in $S \leftrightarrow \sigma(i) \in \omega_i$—in other words, $E_{\sigma(i)}$ is a *Gödel sentence* for ω_i with respect to the system S. [Informally, this means that for any r.e. set, not only is there a Gödel sentence for the set (with respect to S), but "given" any r.e. set, we can "find" such a Gödel sentence.] Another method of proving Theorem 3 (closer to the method we used in earlier chapters to construct Gödel sentences) is provided in Exercise 3 below. A stronger result will be proved in Chapter 12.

From Theorem 3 and the corollary to Theorem 2 we have:

Theorem 4. *If* S *is any consistent axiomatizable extension of* (R), *then* S *is sententially generative. In particular, the systems* $(R), (Q)$ *and P.A. (if consistent) are sententially generative.*

We remark that neither the proof of Theorem 3 nor the proof of Theorem 4 depended on a previously constructed generative r.e. set.

Exercise 3. By the iteration theorem, there is a recursive function $k(y)$ such that for all i, $\omega_{k(i)} = \omega_i*$ $(\omega_i* = d^{-1}(\omega_i))$, where $d(x)$ is the diagonal function). Now suppose $g(x)$ is a recursive function such that for all i, $E_{g(i)}(v_1)$ represents ω_i in S. Show that if we take $r(gkx, gkx)$ for $\sigma(x)$, the conclusion of Theorem 3 holds.

Exercise 4.   a. Show that for any system S, if the set P^* is generative, then the set P is generative.

b. Show that if all r.e. sets are uniformly representable in $\mathcal{S}$, then the set P^* is generative. [This with (a) yields yet another proof that the systems $(R), (Q)$ and P.A. (if consistent) are generative, but it does not show that they are sententially generative].

Exercise 5. Show that the set T of Gödel numbers of the *true* arithmetic sentences and its complement $\tilde{T}$ are both generative sets.

§4. Generativity and Incompleteness.

We now turn to some "effective" analogues of Theorems 7 and 8 of Chapter 2. The three theorems that follow are essentially Kleene's generalized forms of Gödel's theorem (1943), though our terminology and presentation are somewhat different.

We let C be Post's r.e. generative set. We recall that the identity function is a generative function for C.

Theorem K_1. *For any consistent axiomatizable system $\mathcal{S}$, if $H(v_1)$ is a formula that represents C in $\mathcal{S}$ and ω_a is the set represented by $\sim H(v_1)$, then $H(\bar{a})$ is undecidable in $\mathcal{S}$. [Alternatively, and this is closer to Kleene's formulation, if the negation of $H(v_1)$ represents C and ω_b is the set represented by $H(v_1)$, then $H(\bar{b})$ is undecidable in $\mathcal{S}$.]*

Theorem K_2. *For any arithmetically correct axiomatizable system $\mathcal{S}$, if $H(v_1)$ expresses the set $\tilde{C}$ and $H(v_1)$ represents ω_a in $\mathcal{S}$, then $H(\bar{a})$ is a true sentence not provable in $\mathcal{S}$.*

Theorem K_3. *Suppose that $\mathcal{S}$ is an axiomatizable system and that $F(v_1, v_2)$ is a formula that enumerates the set C in $\mathcal{S}$. Let ω_a be the set represented in $\mathcal{S}$ by the formula $\forall v_2 \sim F(v_1, v_2)$. Then:*

1. *If $\mathcal{S}$ is simply consistent, then the sentence $\forall v_2 \sim F(\bar{a}, v_2)$ is not provable in $\mathcal{S}$.*
2. *If $\mathcal{S}$ is ω-consistent, then the sentence $\forall v_2 \sim F(\bar{a}, v_2)$ is not refutable in $\mathcal{S}$.*

Proofs. (K_1) Assume hypothesis. By the assumption of consistency, ω_a is disjoint from C. Therefore, $a \notin \omega_a$ and $a \notin C$. Hence $H(\bar{a})$ is neither refutable nor provable in $\mathcal{S}$. [Alternatively, if $\sim H(v_1)$ represents C and $H(v_1)$ represents ω_b, then ω_b is disjoint from C. Hence b is outside both ω_b and C, and $H(\bar{b})$ is neither provable nor refutable in $\mathcal{S}$.]

(K_2) Assume hypothesis. Since S is correct, the set represented by $H(v_1)$ is a subset of the set expressed by $H(v_1)$; hence $\omega_a \subseteq \tilde{C}$. Thus ω_a is disjoint from C, so a is outside both ω_a and C. Since $a \in \tilde{C}$, $H(\bar{a})$ is true, and since $a \notin \omega_a$, $H(\bar{a})$ is not provable in S (because $H(v_1)$ represents ω_a). So, $H(\bar{a})$ is true but not provable in S.

(K_3) Assume hypothesis.

1. Suppose $\forall v_2 \sim F(\bar{a}, v_2)$ is provable in S. Then $a \in \omega_a$ (because $\forall v_2 \sim F(v_1, v_2)$ represents ω_a), and $a \in C$. Hence for some n, $F(\bar{a}, \bar{n})$ is provable (because $F(v_1, v_2)$ enumerates C). Hence $\exists v_2 F(\bar{a}, v_2)$ is provable—i.e., $\sim \forall v_2 \sim F(\bar{a}, v_2)$ is provable, and S is inconsistent. So if S is consistent, $\forall v_2 \sim F(\bar{a}, v_2)$ is not provable.

2. Suppose S is ω-consistent. Then S is simply consistent; hence $\forall v_2 \sim F(\bar{a}, v_2)$ is not provable (by (1)). Therefore $a \notin \omega_a$ and $a \notin C$ and, therefore, for all n, $F(\bar{a}, \bar{n})$ is refutable (because $F(v_1, v_2)$ enumerates C). Then by ω-consistency, the sentence $\sim \forall v_2 \sim F(\bar{a}, v_2)$ is not provable. [Alternatively, since $F(v_1, v_2)$ enumerates C, then by the ω-consistency lemma of §2, Ch. 0, the formula $\sim \forall v_2 \sim F(v_1, v_2)$ represents C, and so the sentence $\forall v_2 \sim F(\bar{a}, v_2)$ is undecidable by (K_1).]

Discussion. The theorems above provide perfectly good methods of finding an undecidable sentence, say, of P.A. An interesting fact about their proofs is that the diagonal function $d(x)$ is nowhere employed! In fact, no diagonalization *within* the system (say P.A.) is used; rather, the diagonalization (in a more general sense) occurs *outside* the system in the key definition of the set C (we identified the variables x and y in the relation $x \in \omega_y$, and this might be regarded as a "diagonalization" in a more abstract sense). The set C, of course, plays the role in the above proofs that P^* played in our earlier incompleteness proofs.

Neither of the above theorems yields the incompleteness of P.A. under the assumption of *simple* consistency. In the next chapter, we will study Kleene's *symmetric* form of Gödel's theorem; this, like Rosser's argument, establishes the incompleteness of P.A. under the assumption of simple consistency, but it does not use the diagonal function $d(x)$; the diagonalization again occurs *outside* the system.

Chapter V

Double Generativity and Complete Effective Inseparability

I. Complete Effective Inseparability

§1. Complete Effective Inseparability. A disjoint pair (A_1, A_2) is by definition recursively inseparable if no recursive superset of A_1 is disjoint from A_2. This is equivalent to saying that for any disjoint r.e. supersets ω_i and ω_j of A_1 and A_2, the set ω_i is not the complement of ω_j—in other words, there is a number n outside both ω_i and ω_j. The disjoint pair (A_1, A_2) is called *effectively inseparable*—abbreviated E.I.—if there is a recursive function $\delta(x, y)$—called an *E.I. function* for (A_1, A_2)—such that for any numbers i and j such that $A_1 \subseteq \omega_i$ and $A_2 \subseteq \omega_j$ with ω_i being disjoint from ω_j, the number $\delta(i,j)$ is outside both ω_i and ω_j.

We shall call a disjoint pair (A_1, A_2) *completely* E.I. if there is a recursive function $\delta(x, y)$—which we call a complete E.I. function for (A_1, A_2)—such that for any numbers i and j, if $A_1 \subseteq \omega_i$ and $A_2 \subseteq \omega_j$, then $\delta(i,j) \in \omega_i \leftrightarrow \delta(i,j) \in \omega_j$ (in other words, $\delta(i,j)$ is either inside or outside both sets ω_i and ω_j.). [If ω_i and ω_j happen to be disjoint, then, of course, $\delta(i,j)$ is outside both ω_i and ω_j, so any complete E.I. function for (A_1, A_2) is also an E.I. function for (A_1, A_2).] In a later chapter, we will prove the non-trivial fact that if (A_1, A_2) is E.I. and A_1 and A_2 are both r.e., then (A_1, A_2) is completely E.I. [The proof of this uses the result known as the *double recursion theorem*, which we will study in Chapter 9.]

Effective inseparability has been well studied in the literature. Complete effective inseparability will play a more prominent role in this volume—especially in the next few chapters.

Proposition 1.

(1) *If* (A_1, A_2) *is completely E.I., then so is* (A_2, A_1)—*in fact, if* $\delta(x, y)$ *is a complete E.I. function for* (A_1, A_2), *then* $\delta(y, x)$ *is a complete E.I. function for* (A_2, A_1).

(2) *If* (A_1, A_2) *is completely E.I. under* $\delta(x, y)$ *and additionally* $A_1 \subseteq A_1'$, $A_2 \subseteq A_2'$ *and* A_1' *are disjoint from* A_2', *then* (A_1', A_2') *is completely E.I. under* $\delta(x, y)$.

Proof. Obvious.

§2. Kleene's Construction.

There are many ways to construct a completely E.I. pair of r.e. sets and we shall look at several. We first turn to Kleene's method.

We will call a recursive function $g(x, y)$ a *Kleene* function for a disjoint pair (A_1, A_2) if for all x and y

1. $g(x, y) \in \omega_y - \omega_x \Rightarrow g(x, y) \in A_1$,
2. $g(x, y) \in \omega_x - \omega_y \Rightarrow g(x, y) \in A_2$.

Proposition 2. *If* $g(x, y)$ *is a Kleene function for* (A_1, A_2), *then* (A_1, A_2) *is completely E.I. under* $g(x, y)$.

Proof. Suppose $g(x, y)$ is a Kleene function for (A_1, A_2). Now suppose x and y are such that $A_1 \subseteq \omega_x$ and $A_2 \subset \omega_y$. Then

$$g(x, y) \in \omega_y - \omega_x \Rightarrow g(x, y) \in A_1 \Rightarrow g(x, y) \in \omega_x,$$

but no number in $\omega_y - \omega_x$ can be in ω_x. Therefore, $g(x, y) \notin \omega_y - \omega_x$. By a symmetric argument, $g(x, y) \notin \omega_x - \omega_y$. Therefore,

$$g(x, y) \in \omega_x \leftrightarrow g(x, y) \in \omega_y.$$

We shall call a disjoint pair (A_1, A_2) a *Kleene* pair if it has a Kleene function. By Proposition 2, every Kleene pair is completely E.I.

Theorem 1. *There exists a Kleene pair* (K_1, K_2) *of r.e. sets.*

Proof. We use the recursive pairing function $J(x, y)$ and its inverse functions $K(x)$ and $L(x)$, and let Π_1 be the set of all numbers $J(x, y)$ such that $J(x, y) \in \omega_y$. We also let Π_2 be the set of all numbers $J(x, y)$ such that $J(x, y) \in \omega_x$. [Equivalently, Π_1 is the set of all x such that $x \in \omega_{Lx}$ and Π_2 is the set of all x such that $x \in \omega_{Kx}$.]

Obviously

$$J(x,y) \in \omega_y - \omega_x \leftrightarrow J(x,y) \in \Pi_1 - \Pi_2$$

$$J(x,y) \in \omega_x - \omega_y \leftrightarrow J(x,y) \in \Pi_2 - \Pi_1$$

so $J(x,y)$ is a Kleene function for $(\Pi_1 - \Pi_2, \Pi_2 - \Pi_1)$. Since the sets Π_1 and Π_2 are both r.e., by the separation principle (Theorem 10, Chapter 0), there are disjoint r.e. supersets K_1 and K_2 of $\Pi_1 - \Pi_2$ and $\Pi_2 - \Pi_1$ respectively. [For definiteness, we can take K_1 to be the set of all x such that $x \in \omega_{Lx}$ before $x \in \omega_{Kx}$ and take K_2 to be the set of all x such that $x \in \omega_{Kx}$ before $x \in \omega_{Lx}$—i.e.

$$K_1 = x : B(x, Lx, Kx)$$

$$K_2 = x : B(x, Kx, Lx),$$

where $B(x,y,z)$ is the r.e. relation of Theorem 3, Chapter 3.] Since $J(x,y)$ is a Kleene function for the smaller pair ($\Pi_1 - \Pi_2$ and $\Pi_2 - \Pi_1$), then, of course, it is a Kleene function for the larger pair (K_1, K_2).

From Theorem 1 and Proposition 2 we have:

Theorem 2. *There is a completely E.I. pair (K_1, K_2) of r.e. sets.*

§3. Kleene's Symmetric form of Gödel's Theorem.

The following theorem is a "constructive" analogue of Theorem 15, Chapter 2.

Theorem 3. *(Kleene's Symmetric Form of Gödel's Theorem) Suppose S is a consistent axiomatizable system in which the pair (K_1, K_2) is strongly separable. Then for any formula $H(v_1)$ which strongly separates K_1 from K_2 in S, if ω_a is the set represented in S by $H(v_1)$, and ω_b is the set represented in S by $\sim H(v_1)$, then $H(\bar{c})$ is an undecidable sentence of S, where c is the number $J(a,b)$.*

Proof. Assume hypothesis. Then $K_1 \subseteq \omega_a$ and $K_2 \subseteq \omega_b$. Since S is consistent, the sets ω_a and ω_b are disjoint. Therefore,

$$J(a,b) \notin \omega_a \cup \omega_b$$

(because $J(x,y)$ is a Kleene function for (K_1, K_2). Hence $J(x,y)$ is a complete E.I. function for (K_1, K_2) and an E.I. function for (K_1, K_2)). We let $c = J(a,b)$. Since $c \notin \omega_a$, $H(\bar{c})$ is not provable in S and since $c \notin \omega_b$, $H(\bar{c})$ is not refutable in S. Therefore $H(\bar{c})$ is undecidable in S.

Remark. In the above proof, the pair (K_1, K_2) plays an analogous role to the pair (R^*, P^*) of Rosser's proof. Of course, the sets K_1 and K_2 were defined quite independently of the system $\mathcal{S}$, and the diagonal function $d(x)$ was not used.

§4. Reducibility and Semi-reducibility.

We shall say that an ordered pair (A_1, A_2) is *reducible* to an ordered pair (B_1, B_2) if there is a recursive function $f(x)$ which simultaneously reduces A_1 to B_1 and A_2 to B_2. This means that $A_1 = f^{-1}(B_1)$ and $A_2 = f^{-1}(B_2)$, or equivalently that for every number x:

(1) $x \in A_1 \Rightarrow f(x) \in B_1$,
(2) $x \in A_2 \Rightarrow f(x) \in B_2$,
(3) $x \notin A_1 \cup A_2 \Rightarrow f(x) \notin B_1 \cup B_2$.

Such a function $f(x)$ we call a *reduction* of (A_1, A_2) to (B_1, B_2); we also say that (A_1, A_2) is reducible to (B_1, B_2) under $f(x)$.

We call a recursive function $f(x)$ a *semi-reduction* of (A_1, A_2) to (B_1, B_2) if for all x, conditions (1) and (2) above hold (but not necessarily (3)). This means that for all $x \in A_1$, $f(x) \in B_1$ and for all $x \in A_2$ and $f(x) \in B_2$—in other words, $A_1 \subseteq f^{-1}(B_1)$ and $A_2 \subseteq f^{-1}(B_2)$.

The following two lemmas will have several applications.

Lemma A_1. *If (A_1, A_2) is a Kleene pair and if (A_1, A_2) is semi-reducible to (B_1, B_2), and B_1 and B_2 are disjoint, then (B_1, B_2) is a Kleene pair. More specifically, if $g(x, y)$ is a Kleene function for (A_1, A_2), and $f(x)$ is a semi-reduction of (A_1, A_2) to (B_1, B_2), and if B_1 and B_2 are disjoint, then for any iterative function $t(y)$ for the relation $f(x) \in \omega_y$, the function $f(g(tx, ty))$ is a Kleene function for (B_1, B_2).*

Proof. Assume hypothesis.

1. Suppose $fg(tx, ty) \in \omega_y - \omega_x$. Then $g(tx, ty) \in f^{-1}(\omega_y - \omega_x)$. But $f^{-1}(\omega_y - \omega_x) = f^{-1}(\omega_y) - f^{-1}(\omega_x) = \omega_{t(y)} - \omega_{t(x)}$. So $g(t(x), t(y)) \in \omega_{t(y)} - \omega_{t(x)}$. Hence $g(t(x), t(y)) \in A_1$ (since $g(x, y)$ is a Kleene function for (A_1, A_2)). Therefore,

$$fg(tx, ty) \in B_1$$

(because $f(x)$ is a semi-reduction of (A_1, A_2) to (B_1, B_2)).

2. By a similar argument, if $fg(tx, ty) \in \omega_x - \omega_y$, then

$$fg(tx, ty)) \in B_2,$$

and so $fg(tx, ty)$ is a Kleene function for (B_1, B_2).

Lemma A_2. *Let (A, B) be a disjoint pair of sets, and $f(x, y)$ and $g(x, y)$ be recursive functions such that for all i, j and x:*

(1) $x \in \omega_j - \omega_i \Rightarrow f(g(i, j), x) \in A$,
(2) $x \in \omega_i - \omega_j \Rightarrow f(g(i, j), x) \in B$.

    Let $A' = x : f(x, x) \in A$, $B' = x : f(x, x) \in B$ and $t(y)$ be any iterative function for the relation $f(x, x) \in \omega_y$. Then:

(a) $g(x, y)$ *is a Kleene function for* (A', B'),
(b) $f(g(tx, ty), g(tx, ty))$ *is a Kleene function for* (A, B).

Proof. Assume hypothesis.

1. Substituting $g(i, j)$ for x in (1) we get:

$$g(i, j) \in \omega_j - \omega_i \Rightarrow f(g(i, j), g(i, j)) \in A \Rightarrow g(i, j) \in A'.$$

Similarly, from (2) we get:

$$g(i, j) \in \omega_i - \omega_j \Rightarrow g(i, j) \in B',$$

and so $g(x, y)$ is a Kleene function for (A', B').
2. This follows from (a) and Lemma A_1, since $f(x, x)$ is a semi-reduction (in fact a reduction) of (A', B') to (A, B).

§5. Completely E.I. Systems.

We continue to let P and R be the set of Gödel numbers of the provable and refutable formulas of $\mathcal{S}$; we shall refer to P and R as the *nuclei* of $\mathcal{S}$. $\mathcal{S}$ is called an E.I. system if the pair (P, R) is E.I., and call $\mathcal{S}$ *completely* E.I. if the pair (P, R) is completely E.I. We continue to use the representation function $r(x, y)$ ($r(x, y)$ is the Gödel number of $E_x[\bar{y}]$).

Lemma. *If (A_1, A_2) is strongly separable in $\mathcal{S}$, then (A_1, A_2) is semi-reducible to (P, R).*

Proof. Let h be the Gödel number of a formula $H(v_1)$ which strongly separates A_1 from A_2 in $\mathcal{S}$. Then the recursive function $r(h, x)$ (as a function of x) is a semi-reduction of (A_1, A_2) to (P, R).

Note. If $H(v_1)$ *exactly* separates A_1 from A_2 in $\mathcal{S}$, then the function $r(h, x)$ is a reduction of (A_1, A_2) to (P, R).

From the above lemma, Lemma A_1 and Proposition 2, we get:

Theorem 4. *If some Kleene pair is strongly separable in $\mathcal{S}$ and $\mathcal{S}$ is consistent, then its nuclei P and R are completely E.I.—in fact, (P, R) is a Kleene pair.*

Theorem 5. *Every consistent Rosser system for sets is completely E.I.*

Proof. If $\mathcal{S}$ is a consistent Rosser system for sets, then the Kleene pair (K_1, K_2) of r.e. sets is strongly separable in $\mathcal{S}$; hence $\mathcal{S}$ is completely E.I. by Theorem 4.

Corollary. *Every consistent extension of (R) is completely E.I. In particular, the systems $(R), (Q)$, and P.A. (if consistent) are completely E.I.*

Remark. The above proof of the corollary utilizes a Kleene pair of r.e. sets previously constructed.

In the next section, we will obtain a strengthening of the above corollary whose proof does not utilize a previously constructed Kleene pair of r.e. sets and which provides another proof of the existence of such a pair.

§6. Effective Rosser Systems.

We shall say that $\mathcal{S}$ is *effectively* a Rosser system for sets if there is a recursive function $\pi(x, y)$—which we call a *Rosser function* for $\mathcal{S}$—such that for any numbers i and j, the number $\pi(i, j)$ is the Gödel number of a formula $H(v_1)$ which strongly separates $\omega_i - \omega_j$ from $\omega_j - \omega_i$ in $\mathcal{S}$.

Theorem 6. *Suppose $\mathcal{S}$ is a consistent effective Rosser system for sets. Then*

(a) *The pair (P^*, R^*) is completely E.I.—in fact, it is a Kleene pair.*

(b) *The pair (P, R) is also a Kleene pair—moreover, there is a Kleene function $\delta(x, y)$ for (P, R) such that for all numbers i and j, the number $\delta(i, j)$ is the Gödel number of a sentence.*

Proof. Suppose S is consistent and that $\pi(x, y)$ is a Rosser function for S. Let $\pi'(x, y) = \pi(y, x)$. Then for each i and j, $\pi'(i, j)$ is the Gödel number of a formula that strongly separates $\omega_j-\omega_i$ from $\omega_i-\omega_j$ in S. Hence for all i, j and x

1. $x \in \omega_j - \omega_i \Rightarrow r(\pi'(i, j), x) \in P$,
2. $x \in \omega_i - \omega_j \Rightarrow r(\pi'(j, i), x) \in R$.

    a. It then follows from (a) of Lemma A_2 (taking r for f and π' for g) that $\pi'(x, y)$ is a Kleene function for (P^*, R^*) (because $P^* = x : r(x, x) \in P$ and $R^* = x : r(x, x) \in R$, since $r(x, x)$ is the diagonal function $d(x)$).

    b. We now take any iterative function $t(y)$ for the relation $r(x, x) \in \omega_y$. Then by using (b) of Lemma A_2, we find that $r(\pi'(tx, ty), \pi'(tx, ty))$ is a Kleene function for (P, R). Thus $d(\pi'(tx, ty))$ is a Kleene function for (P, R). Since for any numbers a and b, $\pi'(a, b)$ is the Gödel number of a formula with v_1 as the only free variable and $d\pi'(a, b)$ is the Gödel number of a sentence. Hence for all x and y, $d\pi'(tx, ty)$ is a sentence-number (Gödel number of a sentence). We thus take $\delta(x, y)$ to be $d\pi'(tx, ty)$.

Remark. The fact that (P^*, R^*) is a Kleene pair is a stronger conclusion than (P, R) being a Kleene pair because for any system S with nuclei P and R, if the pair (P^*, R^*) is a Kleene pair, then so is (P, R), because (P^*, R^*) is reducible to (P, R) under $r(x, x)$.

    Theorem 6 has an interesting consequence.

Theorem 7. *Suppose S is effectively a Rosser system for sets and S is consistent. Then there is a recursive function $\delta(x, y)$ such that given any consistent axiomatizable extension S' of S and any numbers i and j, such that ω_i is the set of Gödel numbers of the provable formulas of S' and ω_j is the set of Gödel numbers of the refutable formulas of S', $\delta(i, j)$ is the Gödel number of a sentence undecidable in S'.*

Proof. Exercise. [A stronger version of this theorem will be proved in a later chapter.]

Theorem 8. *If S is a Rosser system for binary relations, then S is effectively a Rosser system for sets.*

Proof. Suppose $\mathcal{S}$ is a Rosser system for binary relations. The relations $x \in \omega_{Ky}$ and $x \in \omega_{Ly}$ are r.e.; hence there is a formula $H(v_1, v_2)$ which strongly separates the relation $x \in \omega_{Ky} \wedge x \notin \omega_{Ly}$ from the relation $x \in \omega_{Ly} \wedge x \notin \omega_{Ky}$. Thus $H(v_1, v_2)$ strongly separates the relation $x \in \omega_{Ky} - \omega_{Ly}$ from the relation $x \in \omega_{Ly} - \omega_{Ky}$. Then for any numbers n, i and j, if $n \in \omega_i - \omega_j$, then $n \in \omega_{KJ(i,j)} - \omega_{LJ(i,j)}$; hence $H[\overline{n}, \overline{J(i,j)}]$ is provable in $\mathcal{S}$. Similarly $n \in \omega_j - \omega_i$ implies that $H[\overline{n}, \overline{J(i,j)}]$ is refutable in $\mathcal{S}$. So for any numbers i and j, the formula $H[v_1, \overline{J(i,j)}]$ (i.e. the formula $\forall v_2(v_2 = \overline{J(i,j)} \supset H(v_1, v_2))$) strong separates $\omega_i - \omega_j$ from $\omega_j - \omega_i$. We let $\pi(i,j)$ be the Gödel number of $H[v_1, \overline{J(i,j)}]$. The function $\pi(x, y)$ is recursive and is a Rosser function for $\mathcal{S}$.

Remark. We know that the system (Q) (in fact every consistent axiomatizable extension of (R)) is a consistent Rosser system (for sets and relations). Hence by Theorem 8, it is effectively a Rosser system for sets, and so by Theorem 6, its pair (P, R) of nuclei is completely E.I. (in fact a Kleene pair). We thus have a second proof of the existence of a Kleene pair of r.e. sets (which does not use a Kleene pair already constructed).

Exercise 1. Suppose that there is a recursive function $\psi(x, y)$ such that for any *disjoint* r.e. sets ω_i and $\omega_j, \psi(i, j)$ is the Gödel number of a formula that strongly separates (ω_i, ω_j) in $\mathcal{S}$. Prove that $\mathcal{S}$ is effectively a Rosser system for sets.

Exercise 2. Call $\mathcal{S}$ an effective Rosser system for n-ary relations if there is a recursive function $\pi(x, y)$ such that for all i and $j, \pi(i, j)$ is the Gödel number of a formula $H(v_1, \ldots, v_n)$ which strongly separates $R_i^n - R_j^n$ from $R_j^n - R_i^n$ in $\mathcal{S}$. Prove that for any positive n, if $\mathcal{S}$ is a Rosser system for relations of $n+1$ arguments, then $\mathcal{S}$ is effectively a Rosser system for relations of n arguments. Conclude that if $\mathcal{S}$ is a Rosser system (for sets and relations), then $\mathcal{S}$ is effectively a Rosser system (for sets and relations. [Thus every axiomatizable extension of (R) is effectively a Rosser system (for sets and relations).]

Effectively Exact Rosser Systems. We will say that $\pi(x, y)$ is an *exact* Rosser function (for sets) for $\mathcal{S}$ if for any *disjoint* r.e. sets ω_i and ω_j, $\pi(i, j)$ is the Gödel number of a formula that *exactly* separates (ω_i, ω_j) in $\mathcal{S}$. We shall say also that $\mathcal{S}$ is effectively an exact Rosser system (for sets) if there is an exact Rosser function for $\mathcal{S}$.

The following result will be needed in a later chapter.

Theorem 8.1. *If S is an exact Rosser system for binary relations (i.e. if any two disjoint r.e. binary relations are exactly separable in S), then S is effectively an exact Rosser system for sets.*

Proof. Suppose S is an exact Rosser system for binary relations. Then the relation "$x \in \omega_{Ky}$ before $x \in \omega_{Ly}$" is exactly separated from the relation "$x \in \omega_{Ly}$ before $x \in \omega_{Ky}$" by some formula $H(v_1, v_2)$. Then (by an argument similar to part of the proof of Theorem 8) for any numbers i and j, the formula $H[v_1, \overline{J(i,j)}]$ exactly separates $x : x \in \omega_i$ before $x \in \omega_j$ from $x : x \in \omega_j$ before $x \in \omega_i$. If ω_i and ω_j are disjoint, then the above two sets are ω_i and ω_j respectively. We thus let $\pi(i,j)$ be the Gödel number of $H[v_1, \overline{J(i,j)}]$, and now $\pi(x,y)$ is an exact Rosser function for S.

II.   Double Universality

§7.   Double Universality.

We shall call a disjoint pair (B_1, B_2) *doubly universal*—abbreviated D.U.—if every disjoint pair (A_1, A_2) of r.e. sets is reducible to (B_1, B_2). We shall say that a disjoint pair (B_1, B_2) is *semi-D.U.* if every disjoint pair of r.e. sets is semi-reducible to (B_1, B_2). [In the next chapter, we will prove the fundamental result that every semi-D.U. pair of r.e. sets is D.U. This will afford an alternative proof to that of Shepherdson that every consistent axiomatizable extension of (R) is an exact Rosser system for sets; the two proofs will generalize in different directions.]

From Theorem 1 and Lemma A_1 we at once have

Theorem 9. *If (A_1, A_2) is semi-D.U., then (A_1, A_2) is a Kleene pair (and, hence, is completely E.I.)*

We will call a pair (B_1, B_2) (not necessarily disjoint) D.U.$^+$ if every pair of r.e. sets (whether disjoint or not) is reducible to (B_1, B_2).

We now wish to construct a pair (V_1, V_2) of r.e. sets such that (V_1, V_2) is D.U.$^+$ and we wish to construct a (disjoint) D.U. pair (U_1, U_2) of r.e. sets.

We take V_1 to be the set of all numbers $J(J(x,y), z)$ such that $z \in \omega_y$ and take V_2 to be the set of all numbers $J(J(x,y), z)$ such that $z \in \omega_x$. The pair (V_1, V_2) is D.U.$^+$ because for any numbers i and j, the pair (ω_i, ω_j) is reduced to (V_1, V_2) by the function $J(J(j,i), x)$.

We take U_1 to be the set of all numbers $J(J(x,y), z)$ such that $z \in \omega_y$ before $z \in \omega_x$, and take U_2 to be the set of all numbers

$J(J(x, y), z)$ such that $z \in \omega_x$ before $z \in \omega_y$.

Now suppose ω_i and ω_j are disjoint. Then for any x

$$x \in \omega_i \leftrightarrow (x \in \omega_i \text{ before } x \in \omega_j) \leftrightarrow J(J(j, i), x) \in U_1.$$

Also

$$x \in \omega_j \leftrightarrow (x \in \omega_j \text{ before } x \in \omega_i) \leftrightarrow J(J(j, i), x) \in U_2.$$

Therefore $J(J(j, i), x)$ reduces (ω_i, ω_j) to (U_1, U_2) and so (U_1, U_2) is doubly universal. And so we have proved:

Theorem 10.

(a) *The pair (V_1, V_2) of r.e. sets is D.U.*[+]
(b) *The (disjoint) pair (U_1, U_2) of r.e. sets is D.U.*

The Pair (U_1', U_2'). We let $U_1' = x : J(x, x) \in U_1$ and we let

$$U_2' = x : J(x, x) \in U_2.$$

For any x, i and j we have

$$x \in \omega_j - \omega_i \Rightarrow (x \in \omega_j \text{ before } x \in \omega_i) \Rightarrow J(J(i, j), x) \in U_1,$$

and similarly

$$x \in \omega_i - \omega_j \Rightarrow J(J(i, j), x) \in U_2.$$

Then by (a) of Lemma A_2 (taking J for f and J for g), $J(x, y)$ is a Kleene function for (U_1', U_2'). This is not too surprising considering that (U_1', U_2') is the very pair (K_1, K_2)! (cf. Exercise 3 below).

Exercise 3.

1. Show that for any numbers x, y, z and w, if $w = J(J(x, y), z)$ then $x = KKw$ and $y = LKw$ and $z = Lw$.
2. Show from this that U_1 is the set of all numbers $J(x, y)$ such that $y \in \omega_{Lx}$ before $y \in \omega_{Kx}$, and U_2 is the set of all numbers $J(x, y)$ such that $y \in \omega_{Kx}$ before $y \in \omega_{Lx}$.
3. Show from this that $U_1' = K_1$ and $U_2' = K_2$.

Exercise 4. Call a disjoint pair (A, B) *uniformly* D.U. if there is a recursive function $f(x, y, z)$ such that for all numbers i and j such that ω_i and ω_j are disjoint, $f(i, j, x)$ reduces (ω_i, ω_j) to (A, B).

Prove that every D.U. pair is uniformly D.U.

Exercise 5. Show that if (A, B) is uniformly D.U., then there are recursive functions $f(x, y)$ and $g(x, y)$ such that for all numbers i

and j for which ω_i and ω_j are disjoint, $g(f(i,j),x)$ reduces (ω_i,ω_j) to (A,B).

III. Double Generativity

§8. Doubly Generative Pairs.

§8. **Doubly Generative Pairs.** We shall call a disjoint pair (A_1, A_2) *doubly generative*—abbreviated D.G.—if there is a recursive function $\phi(x,y)$—called a D.G. function for (A_1, A_2)—such that for any numbers i and j for which ω_i and ω_j are disjoint, the following two conditions both hold:

  1. $\phi(i,j) \in A_1 \leftrightarrow \phi(i,j) \in \omega_i$,
  2. $\phi(i,j) \in A_2 \leftrightarrow \phi(i,j) \in \omega_j$.

Proposition 3. *If (A_1, A_2) is doubly generative, then A_1 and A_2 are each generative.*

Proof. Let $\phi(x,y)$ be a D.G. function for (A_1, A_2). Let a be any index of the empty set. Then for any i, ω_i is disjoint from ω_a and, hence, by (1),

$$\phi(i,a) \in A_1 \leftrightarrow \phi(i,a) \in \omega_i.$$

Therefore, $\phi(x,a)$ is a generative function for A_1. Similarly, by (2), $\phi(a,x)$ is a generative function for A_2.

Proposition 4. *If $\phi(x,y)$ is a D.G. function for (A_1, A_2), then $\phi(y,x)$ is a D.G. function for (A_2, A_1).*

Proof. Obvious.

Exercise 6. Show that the pair (K_1, K_2) is D.G. under the function $J(y,x)$.

The next theorem is basic.

Theorem 11. *If (A_1, A_2) is completely E.I. and A_1 and A_2 are both r.e., then (A_1, A_2) is D.G. Moreover, if A_1 and A_2 are both r.e., then for any complete E.I. function $\delta(x,y)$ for the pair (A_2, A_1), there are recursive functions $t_1(x)$ and $t_2(x)$ such that $\delta(t_2(x), t_1(y))$ is a D.G. function for (A_1, A_2).*

Proof. Suppose $\delta(x,y)$ is a complete E.I. function for (A_2, A_1), and A_2 and A_1 are both r.e. By the iteration theorem, there are recursive functions $t_1(y)$ and $t_2(y)$ such that for all n,

$$\omega_{t_1(n)} = \omega_n \cup A_1$$

and

$$\omega_{t_2(n)} = \omega_n \cup A_2$$

(we take $t_1(y)$ to be an iteration function for the relation

$$x \in \omega_y \vee x \in A_1;$$

$t_2(y)$ is an iteration function for the relation $x \in \omega_y \vee x \in A_2$). We show that the function $\delta(t_2(x), t_1(y))$ is a D.G. function for (A_1, A_2).

Take any numbers i and j and let $k = \delta(t_2(i), t_1(j))$. Since we know that $A_2 \subseteq \omega_{t_2(i)}$ and $A_1 \subseteq \omega_{t_1(j)}$ and $\delta(x, y)$ is a complete E.I. function for (A_2, A_1), then

$$\tag{1} k \in \omega_{t_2(i)} \leftrightarrow k \in \omega_{t_1(j)}.$$

We are to show that if ω_i and ω_j are disjoint, then $k \in A_1 \leftrightarrow k \in \omega_i$ and $k \in A_2 \leftrightarrow k \in \omega_j$. So suppose ω_i and ω_j are disjoint. Suppose also that $k \in A_1$ and $k \in \omega_{t_1(j)}$. Then $k \in \omega_{t_2(i)}$. So $k \in A_2 \cup \omega_i$. But $k \notin A_2$ (since A_2 is disjoint from A_1) and, hence, $k \in \omega_i$. Conversely, suppose $k \in \omega_i$. Then $k \in \omega_{t_2(i)}$ (since $\omega_i \subseteq \omega_{t_2(i)}$). Hence $k \in \omega_{t_1(j)}$ and $k \in \omega_j \cup A_1$. But $k \notin \omega_j$ (since ω_j is disjoint from ω_i) and, hence, $k \in A_1$. This proves that $k \in A_1 \leftrightarrow k \in \omega_i$. The proof that $k \in A_2 \leftrightarrow k \in \omega_j$ is symmetric.

Since there exists a completely E.I. pair of r.e. sets, we have:

Corollary 1. *There exists a D.G. pair of r.e. sets.*

Remark. This corollary was established more easily by Exercise 6.

Corollary 2. *If (A_1, A_2) is completely E.I. and A_1 and A_2 are both r.e., then A_1 and A_2 are both completely creative sets.*

Proof. By Theorem 11 and Proposition 3 (cf. §1 of Chapter 4).

Exercise 7. The last corollary can be proved more simply and without appeal to the notion of double generativity: Let $\delta(x, y)$ be a complete E.I. function for (A_1, A_2) and let A_1 and A_2 be r.e. Let a be any index of A_1 and let $t(y)$ be a recursive function such that for all i, $\omega_{t(i)} = \omega_i \cup A_2$. Show that $\delta(a, t(x))$ is a generative function for A_1.

Exercise 8. Prove that if (A_1, A_2) is E.I. (not necessarily completely E.I.) and A_1 and A_2 are both r.e., then A_1 and A_2 are both creative sets [Myhill].

A particularly important consequence of Theorem 11 for purposes of the next chapter is the following theorem.

Theorem 12. *If* (A_1, A_2) *is semi-D.U. and* A_1 *and* A_2 *are both r.e., then* (A_1, A_2) *is D.G.*

Proof. If (A_1, A_2) is semi-D.U., then it is completely E.I. by Theorem 9. If also A_1 and A_2 are both r.e., then (A_1, A_2) is D.G. by Theorem 11.

*Semi-D.G. Pairs. In the next chapter, we will prove a result that implies that every D.G. pair (A_1, A_2) is completely E.I. (whether A_1 and A_2 are r.e. or not). However, this can be shown by a more direct argument which also establishes something stronger: Call a recursive function $g(x, y)$ a *semi-D.G.* function for a disjoint pair (A_1, A_2) if for all numbers i and j, such that ω_i is disjoint from ω_j, the following two implications hold:

1. $g(i, j) \in \omega_i \Rightarrow g(i, j) \in A_1$,
2. $g(i, j) \in \omega_j \Rightarrow g(i, j) \in A_2$.

If we replace the implications by equivalences, we get the definition of a D.G. function, so every D.G. pair is semi-D.G. (i.e., has a semi-D.G. function).

Theorem 13. *If* (A_1, A_2) *is semi-D.G., then* (A_1, A_2) *is a Kleene pair.*

Proof. (Sketch) Suppose $g(x, y)$ is a semi-D.G. function for (A_1, A_2). Let $\sigma(x, y)$ be the effective separation function of Theorem 4, Chapter 3. Then the function $g(\sigma(x, y), \sigma(y, x))$ can be shown to be a Kleene function for the pair (A_2, A_1).

Corollary. *Every D.G. pair is completely E.I. (in fact is a Kleene pair).*

Exercise 9. Complete the proof of Theorem 13.

§9.   Reducibility.

Theorem 14.

(a) *If (A_1, A_2) is D.G., (A_1, A_2) is reducible to (B_1, B_2), and B_1 and B_2 are disjoint, then (B_1, B_2) is D.G.*

(b) *Every D.U. pair is D.G.*

Proof.

(a) Assume hypothesis. Let $g(x, y)$ be a D.G. function for (A_1, A_2) and let $f(x)$ be a reduction of (A_1, A_2) to (B_1, B_2). Let $t(y)$ be an iteration function for the relation $f(x) \in \omega_y$ (thus we have $\omega_{t(y)} = f^{-1}(\omega_y)$). We show that $fg(tx, ty)$ is a D.G. function for (B_1, B_2). Suppose ω_i and ω_j are disjoint. Then $f^{-1}(\omega_i)$ and $f^{-1}(\omega_j)$ are disjoint, so $\omega_{t(i)}$ and $\omega_{t(j)}$ are disjoint. Therefore,

$$g(t(i), t(j)) \in \omega_{t(i)} \leftrightarrow g(t(i), t(j)) \in A_1$$

(since $g(x, y)$ is a D.G. function for (A_1, A_2)). Therefore

$$fg(t(i), t(j)) \in \omega_i \leftrightarrow g(t(i), t(j)) \in f^{-1}(\omega_i) \leftrightarrow$$
$$g(t(i), t(j)) \in \omega_{t(i)} \leftrightarrow g(t(i), t(j)) \in A_1 \leftrightarrow fg(t(i), t(j)) \in B_1$$

(since $f(x)$ reduces A_1 to B_1). This proves that

$$fg(t(i), t(j)) \in \omega_i \leftrightarrow fg(t(i), t(j)) \in B_1.$$

Similarly, since

$$g(t(i), t(j)) \in \omega_{t(j)} \leftrightarrow g(t(i), t(j)) \in A_2,$$

it follows that

$$fg(t(i), t(j)) \in \omega_j \leftrightarrow fg(t(i), t(j)) \in B_2.$$

(b) This follows from (a) and the existence of a D.G. pair of r.e. sets (Corollary 2 to Theorem 11, or alternatively, Exercise 6).

§10.   Sentential Double Generativity.

Theorem 15. *If S is a consistent axiomatizable Rosser system for sets, then the pair (P, R) of its nuclei is D.G.*

Proof. Assume hypothesis. Then (P, R) is completely E.I. by Theorem 5. Since S is axiomatizable, the sets P and R are both r.e.; hence (P, R) is D.G. by Theorem 11. [Alternatively, the hypothesis implies that (P, R) is a semi-D.U. pair of r.e. sets and is, therefore, D.G. by Theorem 12.]

Sentential Double Generativity. We shall call the system $\mathcal{S}$ *sententially D.G.* if there is a D.G. function $g(x,y)$ for (P,R) with the added property that for any numbers i and j, the number $g(i,j)$ is the Gödel number of a sentence. This implies that for any i and j for which ω_i and ω_j are disjoint, $g(i,j)$ is the Gödel number of a sentence X such that X is provable in $\mathcal{S}$ iff its Gödel number is in ω_i, and X is refutable in $\mathcal{S}$ iff its Gödel number is in ω_j! The systems (R), (Q) and P.A. (if consistent) enjoy this nice property by virtue of the following theorem.

Theorem 16. *If $\mathcal{S}$ is a consistent axiomatizable effective Rosser system for sets, then $\mathcal{S}$ is sententially D.G.*

Proof. Assume hypothesis. Then by (b) of Theorem 6, there is a complete E.I. function (in fact a Kleene function) $\delta(x,y)$ for (P,R) such that for all i and j, $\delta(i,j)$ is the Gödel number of a sentence. Since P and R are both r.e., then by Theorem 11 there are recursive functions $t_1(x)$ and $t_2(x)$ such that the function $\delta(t_1(y),t_2(x))$ is a D.G. function for (P,R). So we take $g(x,y)$ to be $\delta(t_1(y),t_2(x))$. Since $\delta(x,y)$ is always a sentence-number, the same is true of $g(x,y)$.

Corollary. *Every consistent axiomatizable extension of (R) is sententially D.G.*

Remark. A much stronger result will be proved in Chapter 12.

Chapter VI

Universal and Doubly Universal Systems

We now turn to two theorems (Theorems A and B below) that will play a major role in this study. We will give three different proofs of them in the course of this volume, since each proof reveals certain interesting features of its own.

Theorem A. *If* (A_1, A_2) *is semi-D.U. and* A_1 *and* A_2 *are both r.e., then* A_1 *and* A_2 *are both universal sets.*

Theorem B. *If* (A_1, A_2) *is semi-D.U. and* A_1 *and* A_2 *are both r.e., then* (A_1, A_2) *is D.U.*

Of course, Theorem A is a trivial corollary of Theorem B, but our proofs of Theorem A reveal facts not revealed by our proofs of Theorem B.[1]

We give our first proofs in this chapter. After each proof, we establish a metamathematical corollary: Theorem A yields the result of Ehrenfeucht-Feferman [1960] that for any consistent axiomatizable Rosser system S for sets in which all recursive functions of one argument are strongly definable, all r.e. sets are representable in S. Theorem B yields the stronger result of Putnam-Smullyan [1960]— that any such system S is an exact Rosser system for sets. This result is apparently incomparable in strength with Shepherdson's result that any consistent axiomatizable Rosser system for binary relations is an exact Rosser system for sets. Both results, of course, yield different proofs that every consistent axiomatizable extension of (R) is an exact Rosser system for sets.

---

[1] Also, in the next chapter we wil prove a strengthening of Theorem A which does not appear to be derivable from Theorem B.

I.   *Universality*

§1.   **Generativity and Universality.**   We have shown that every universal set is generative. Our first proof of Theorem A will be based on the converse.

Theorem 1. *Every generative set is universal.*[2]

We will, in fact, prove something considerably stronger which will have other applications as well.

Consider a collection C of r.e. sets. Following Smullyan (1963), we will say that a set A is generative *relative to C* if there is a recursive function $\phi(x)$—called a generative function for A *relative to C*—such that for any number i for which $\omega_i \in C$,

$$\phi(i) \in A \leftrightarrow \phi(i) \in \omega_i.$$

What we have previously called *generative* is, thus, generative with respect to the collection of *all* r.e. sets. We will show that a sufficient condition for a set α to be universal is that it be generative relative to the collection of all *recursive* sets. In fact, we will show the more surprising fact that if α is generative relative to the collection consisting of just the two sets N and $\emptyset$, then α is universal!

For any set α and any number x, the condition that

$$x \in N \leftrightarrow x \in \alpha$$

is obviously equivalent to the condition that $x \in \alpha$ (since $x \in N$), and the condition that

$$x \in \emptyset \leftrightarrow x \in \alpha$$

is equivalent to the condition that $x \notin \alpha$ (since $x \notin \emptyset$). Therefore, the following proposition is obvious.

Proposition 1. α *is generative relative to* $\{N, \emptyset\}$ *under* $\phi(x)$ *($\phi(x)$ recursive) iff for any number i*

---

[2] This says that a set whose complement is *completely* productive must be universal. We will later prove Myhill's stronger result that a set whose complement is *productive* must be universal. The proof of this stronger theorem requires a fixed-point argument (the recursion theorem) that we will study later. Though Theorem 1 is weaker than that of Myhill, it is still strong enough to yield Theorem A (and hence strong enough to establish the Ehrenfeucht-Feferman theorem).

(1) $\omega_i = N \Rightarrow \phi(i) \in \alpha$;
(2) $\omega_i = \emptyset \Rightarrow \phi(i) \notin \alpha$.

Thus, a generative function for α relative to $\{N, \emptyset\}$ is nothing more nor less than a recursive function $\phi(x)$ which maps every index of the set N of all natural numbers to a number inside α and every index of the empty set to a number outside α. Now we need a lemma:

Lemma 1. *For every r.e. set A, there is a recursive function $t(y)$ such that for every number i*

(1) *If $i \in A$, then $\omega_{t(i)} = N$,*
(2) *If $i \notin A$, then $\omega_{t(i)} = \emptyset$.*

Proof. Let A be r.e. Let M be the set of all ordered pairs (x, y) such that $y \in A$. Then M is an r.e. relation and for all x and y

$$M(x, y) \leftrightarrow y \in A.$$

By the iteration theorem, there is a recursive function $t(y)$ such that for all i, $\omega_{t(i)} = x : M(x, i)$. Hence $\omega_{t(i)} = x : i \in A$. Thus, for any x, $x \in \omega_{t(i)} \leftrightarrow i \in A$. If $i \in A$, then for every x, $x \in \omega_{t(i)}$, which means $\omega_{t(i)} = N$. If $i \notin A$, then for every x, $x \notin \omega_{t(i)}$, which means $\omega_{t(i)} = \emptyset$. This proves the lemma.

Now we prove the following strengthening of Theorem 1.

Theorem 1*. *If α is generative relative to $\{N, \emptyset\}$, then α is universal.*

Proof. Let $\phi(x)$ be a generative function for α relative to $\{N, \emptyset\}$. Let A be any r.e. set that we wish to reduce to α. Let $t(y)$ be a recursive function related to A as in Lemma 1. Then for any number i

1. $i \in A \;\; \Rightarrow \omega_{t(i)} = N$ (by Lemma 1),
    $\Rightarrow \phi t i \in \alpha$ (by Proposition 1),
2. $i \notin A \;\; \Rightarrow \omega_{t(i)} = \emptyset$ (by Lemma 1),
    $\Rightarrow \phi t i \notin \alpha$ (by Proposition 1).

By 1 and 2, the recursive function $\phi t(x)$ reduces A to α. This concludes the proof.

Now we prove Theorem A. Theorem 1, of course, follows from Theorem 1*. Now suppose (A_1, A_2) is a semi-D.U. pair of r.e. sets. Then (A_1, A_2) is D.G. (by Theorem 12, Chapter 5); hence A_1 and A_2 are both generative (by Proposition 3, Chapter 5) and so A_1 and A_2 are both universal by Theorem 1.

§2.   The Ehrenfeucht-Feferman Theorem.   Now we consider some metamathematical applications of Theorem A. First we give a lemma.

Lemma 2. *If S is a consistent axiomatizable Rosser system for sets, then some semi-D.U. pair of r.e. sets is exactly separable in S.*

Proof. Assume hypothesis. Take any D.U. pair (A_1, A_2) of r.e. sets. Since S is a Rosser system for sets, then (A_1, A_2) is strongly separable in S by some formula $F(v_1)$. Let B_1 and B_2 be the sets represented in S by $F(v_1)$, $\sim F(v_1)$ respectively. Since S is consistent, B_1 and B_2 are disjoint. Also $A_1 \subseteq B_1$ and $A_2 \subseteq B_2$. Therefore (B_1, B_2) is semi-D.U. (because (A_1, A_2) is semi-D.U., being D.U.). Since S is axiomatizable, the sets B_1 and B_2 are both r.e. Of course, (B_1, B_2) is exactly separated in S by $F(v_1)$.

Theorem A_1. *Suppose S is a consistent axiomatizable Rosser system for sets. Then*

(a) *Some universal set is representable in S.*
(b) *If also all recursive functions of one argument are strongly definable in S, then all r.e. sets are representable in S.*

Proof. Assume hypothesis.

1. By the above lemma, some semi-D.U. pair (B_1, B_2) of r.e. sets is exactly separable in S. By Theorem A, the set B_1 (and also B_2) is universal. So B_1 is a universal set representable in S.
2. Suppose also that all recursive functions of one argument are strongly definable in S. Now take any r.e. set A. Since B_1 is universal, then A is reducible to B_1, so there is a recursive function $f(x)$ such that $A = f^{-1}(B_1)$. By hypothesis, $f(x)$ is strongly definable in S. Then by Th. 11.2, Ch. 0, the set $f^{-1}(B_1)$ is representable in S. Thus A is representable in S.

Statement (b) of Theorem A_1 is the Ehrenfeucht-Feferman Theorem.

II.   Double Universality

§3.   Double Generativity and Double Universality.   The following "double analogue" of Theorem 1 will be the basis for our first proof of Theorem B.

Theorem 2. *If* (A_1, A_2) *is D.G., then* (A_1, A_2) *is D.U.*

Again we will prove something stronger. Let $\mathcal{D}$ be a collection of *ordered pairs* of r.e. sets. A disjoint pair (A_1, A_2) will be said to be doubly generative *relative to* $\mathcal{D}$ if there is a recursive function $g(x, y)$—which we call a D.G. function for (A_1, A_2) relative to $\mathcal{D}$— such that for any numbers i and j, if $(\omega_i, \omega_j) \in \mathcal{D}$, then

$$g(i, j) \in \omega_i \leftrightarrow g(i, j) \in A_1$$

and

$$g(i, j) \in \omega_j \leftrightarrow g(i, j) \in A_2.$$

To say that (A_1, A_2) is doubly generative (as previously defined) is to say that (A_1, A_2) is D.G. relative to the collection of all disjoint pairs of r.e. sets. We will show that a sufficient condition for a pair to be doubly universal is that it be D.G. relative to the collection of all disjoint pairs of *recursive* sets. We will also show that if (A_1, A_2) is D.G. relative to the collection of all *complementary* pairs $(C, \tilde{C})$ of recursive sets, then (A_1, A_2) is semi-D.U. [This fact will have an interesting metamathemtical application.] Actually, we will show stronger versions of both these facts: We let $\mathcal{D}_2$ and $\mathcal{D}_3$ be the following finite collections of disjoint pairs of r.e. sets.

$$\begin{aligned} \mathcal{D}_2 &= \{(N, \emptyset), (\emptyset, N)\}, \\ \mathcal{D}_3 &= \{(N, \emptyset), (\emptyset, N), (\emptyset, \emptyset)\}. \end{aligned}$$

We will prove

Theorem 2*.

(a) *If* (A_1, A_2) *is D.G. relative to* $\mathcal{D}_2$, *then* (A_1, A_2) *is semi-D.U.*
(b) *If* (A_1, A_2) *is D.G. relative to* $\mathcal{D}_3$, *then* (A_1, A_2) *is D.U.*

Of course (b) of Theorem 2* implies Theorem 2. In preparation for the proof, let us consider the following three conditions which may or may not hold for a given recursive function $g(x, y)$ and a given disjoint pair (α_1, α_2) of sets.

C_1 For all i and j, if $\omega_i = N$ and $\omega_j = \emptyset$, then $g(i, j) \in \alpha_1$.
C_2 For all i and j, if $\omega_i = \emptyset$ and $\omega_j = N$, then $g(i, j) \in \alpha_2$.
C_3 For all i and j, if $\omega_i = \emptyset$ and $\omega_j = \emptyset$, then $g(i, j) \notin \alpha_1 \cup \alpha_2$.

Proposition 2.

(1) (α_1, α_2) *is D.G. relative to* $\mathcal{D}_2$ *under* $g(x, y)$ *iff conditions* C_1 *and* C_2 *both hold.*

(2) (α_1, α_2) *is D.G. relative to* $\mathcal{D}_3$ *under* $g(x,y)$ *iff conditions* C_1, C_2 *and* C_3 *all hold.*

We leave the verification of Proposition 2 to the reader. [Again, the crucial point is that

$$(x \in N \leftrightarrow x \in \alpha_1) \leftrightarrow x \in \alpha_1; (x \in \emptyset \leftrightarrow x \in \alpha_1) \leftrightarrow x \notin \alpha_1,$$

and the same with α_2.]

Proof of Theorem 2*:

(a) Suppose that $g(x,y)$ is a D.G. function for (α_1, α_2) relative to $\mathcal{D}_2$. Then conditions C_1 and C_2 both hold. Let (A_1, A_2) be any disjoint pair of r.e. sets that we wish to "semi-reduce" to (α_1, α_2). By Lemma 1, there is a recursive function $t_1(y)$ such that for all $i \in A_1$, $\omega_{t(i)} = N$, and for all $i \notin A_1$, $\omega_{t(i)} = \emptyset$. Again by Lemma 1, there is a recursive function $t_2(y)$ such that for any $i \in A_2$, $\omega_{t_2(i)} = N$ and for any $i \notin A_2, \omega_{t_2(i)} = \emptyset$. We show that the function $g(t_1(x), t_2(x))$ is a semi-reduction of (A_1, A_2) to (α_1, α_2).

   (1) Suppose $i \in A_1$. Then $i \notin A_2$ (since A_1 and A_2 are disjoint) and so $\omega_{t_1(i)} = N$ and $\omega_{t_2(i)} = \emptyset$. Hence by C_1,

   $$g(t_1(i), t_2(i)) \in \alpha_1.$$

   (2) Suppose $i \in A_2$. Then $i \notin A_1$, so $\omega_{t_1(i)} = \emptyset$ and $\omega_{t_2(i)} = N$. Hence by C_2,

   $$g(t_1(i), t_2(i)) \in \alpha_2.$$

Therefore, we see that $g(t_1(x), t_2(x))$ is a semi-reduction of (A_1, A_2) to (α_1, α_2). This proves (a).

(b) Suppose $g(x,y)$ is a D.G. function for (α_1, α_2) relative to $\mathcal{D}_3$. Then condition C_3 also holds. Now take any $i \notin A_1 \cup A_2$. Then $i \notin A_1$ and $i \notin A_2$, so $\omega_{t_1(i)} = \emptyset$ and $\omega_{t_2(i)} = \emptyset$. Then by C_3,

$$g(t_1(i), t_2(i)) \notin \alpha_1 \cup \alpha_2.$$

This, with (1) and (2), implies that the function $g(t_1(x), t_2(x))$ is a reduction of (A_1, A_2) to (α_1, α_2).

Proof of Theorem B: Theorem 2, of course, follows from (b) of Theorem 2*. Now suppose (α_1, α_2) is semi-D.U. and α_1 and α_2 are both r.e. Then (α_1, α_2) is D.G. (by Theorem 12, Chapter 5) and so (α_1, α_2) is D.U. by Theorem 2.

Before turning to the metamethematical applications of Theorem B, let us note some other consequences of Theorem 2.

Corollary (of Theorem 2). *If (A, B) is completely E.I. and A and B are r.e., then (A, B) is D.U.*

Proof. If (A, B) is completely E.I. and A and B are r.e., then (A, B) is D.G. (by Theorem 11, Chapter 5) and, hence, (A, B) is D.U. by Theorem 2.

Theorem 2 also affords another proof that every D.G. pair is a Kleene pair (and, hence, is completely E.I.). If (A, B) is D.G., then it is D.U. (by Theorem 2). Hence it is certainly semi-D.U. and a Kleene pair by Theorem 9, Chapter 5.

We now see that for any disjoint pair (A, B) *of r.e. sets*, the following conditions are all equivalent: (1) (A, B) is a Kleene pair; (2) (A, B) is completely E.I; (3) (A, B) is semi-D.G.; (4) (A, B) is D.G.; (5) (A, B) is semi-D.U.; (6) (A, B) is D.U.

§4. Metamathematical Applications.

Using Theorem B in place of Theorem A, we can obtain the following strengthening of Theorem A_1.

Theorem B_1. *Suppose S is a consistent axiomatizable Rosser system for sets. Then*

(a) *Some D.U. pair is exactly separable in S,*
(b) *If also all recursive functions of one argument are strongly definable in S, then S is an exact Rosser system for sets.*

Proof. Assume hypothesis. By Lemma 2 some semi-D.U. pair (B_1, B_2) of r.e. sets is exactly separable in S.

(a) By Theorem B, the pair (B_1, B_2) is D.U.
(b) Now suppose the additional hypothesis of (b). Let (A_1, A_2) be any disjoint pair of r.e. sets. Since (B_1, B_2) is D.U., then there is a recursive function $f(x)$ such that $A_1 = f^{-1}(B_1)$ and $A_2 = f^{-1}(B_2)$. Then (A_1, A_2) is exactly separable in S (by (2) of Th. 11.2, Ch. 0, since $f(x)$ is strongly definable in S).

Statement (b) of Theorem B_1 is the Putnam-Smullyan Theorem.

Chapter VII

Shepherdson Revisited

Theorems A and B of the last chapter were proved using previous results about generativity, Kleene pairs, complete effective inseparability, and double generativity. Yet the two theorems made no mention of these notions; they referred only to the notions of universality, double universality and semi-double universality. [These three notions, by the way, unlike the four notions mentioned above, were defined without reference to any indexing; they are what we would call *index-free*.] Is it not possible to give more direct proofs of Theorems A and B that do not require all the antecedent machinery of Chapters 4 and 5? We are about to show that it is possible; we will simply transfer Shepherdson's arguments about first-order systems to recursion theory itself. We shall prove some "recursive function-theoretic" analogues of Shepherdson's theorems which will provide new proofs of Theorems A and B (in fact, a strengthening of Theorem A will result).

§1. **Separation Functions.** Let (α, β) be a disjoint pair of sets (not necessarily r.e.). By a *separation* function for (α, β), we shall mean a recursive function $S(x, y, z)$ such that for any r.e. relations $M_1(x, y)$ and $M_2(x, y)$ there is a number h such that for all x and y

(1) $M_1(x, y) \wedge \sim M_2(x, y) \Rightarrow S(h, x, y) \in \alpha$,
(2) $M_2(x, y) \wedge \sim M_1(x, y) \Rightarrow S(h, x, y) \in \beta$.

As an example, if $\mathcal{S}$ is any Rosser system for binary relations, then the function $r_2(x, y, z)$ is a separation function for the pair (P, R) (we recall that $r_2(x, y, z) \underset{\mathrm{df}}{=}$ Gödel number of $E_x[\bar{y}, \bar{z}]$).

We now employ the doubly universal pair (U_1, U_2) of r.e. sets constructed in Chapter 5.

Lemma. *For any r.e. sets A_1 and A_2 there is a number h such that the function $J(h, x)$ is a semi-reduction of $(A_1 - A_2, A_2 - A_1)$ to (U_1, U_2).*

Proof. For any r.e. sets A_1 and A_2, let i be an index of A_1 and j be an index of A_2, and let $h = J(j, i)$. For any x, if $x \in \omega_i - \omega_j$, then $x \in \omega_i$ before $x \in \omega_j$. Hence $J(J(j, i), x) \in U_1$. If $x \in \omega_j - \omega_i$, then $x \in \omega_j$ before $x \in \omega_i$. Hence $J(J(j, i), x) \in U_2$. So $J(h, x)$ is a semi-reduction of $(A_1 - A_2, A_2 - A_1)$ to (U_1, U_2).

Proposition 1. *If (α, β) is semi-D.U. (α, β are not necessarily r.e.), then there is a separation function $S(x, y, z)$ for (α, β).*

Proof. Suppose (α, β) is semi-D.U. Then there is a recursive function $f(x)$ such that $f(x)$ is a semi-reduction of (U_1, U_2) to (α, β). We let $S(x, y, z)$ be the recursive function $fJ(x, J(y, z))$ and we will show that it is a separation function for (α, β).

Take any r.e. relations $M_1(x, y)$ and $M_2(x, y)$. Let A_1 be the set of all numbers $J(x, y)$ such that $M_1(x, y)$ and let A_2 be the set of all numbers $J(x, y)$ such that $M_2(x, y)$. By the above lemma, there is a number h such that $J(h, x)$ is a semi-reduction of $(A_1 - A_2, A_2 - A_1)$ to (U_1, U_2). Therefore, $fJ(h, x)$ is a semi-reduction of

$$(A_1 - A_2, A_2 - A_1)$$

to (α, β). Therefore, for any numbers x and y:

1. $M_1(x, y) \land \sim M_2(x, y) \Rightarrow J(x, y) \in A_1 - A_2 \Rightarrow$
    $fJ(h, J(x, y)) \in \alpha \Rightarrow S(h, x, y) \in \alpha,$
2. $M_2(x, y) \land \sim M_1(x, y) \Rightarrow J(x, y) \in A_2 - A_1 \Rightarrow$
    $fJ(h, J(x, y)) \in \beta \Rightarrow S(h, x, y) \in \beta.$

Theorem 1. *If there is a separation function for $(\alpha, \tilde{\alpha})$ and α is r.e., then α is universal.*

Proof [after Shepherdson]. Suppose $S(x, y, z)$ is a separation function for $(\alpha, \tilde{\alpha})$ and α is r.e. Let A be any r.e. set. We will show that A is reducible to α.

Define $M_1(x, y)$ iff $x \in A$ and define $M_2(x, y)$ iff $S(y, x, y) \in \alpha$. Since A and α are r.e., then the relations $M_1(x, y)$ and $M_2(x, y)$ are r.e. Then there is a number h such that for all x and y

1. $x \in A \wedge S(y,x,y) \notin \alpha \Rightarrow S(h,x,y) \in \alpha$,
2. $S(y,x,y) \in \alpha \wedge x \notin A \Rightarrow S(h,x,y) \notin \alpha$.

Taking h for y, it follows by propositional logic that

$$x \in A \leftrightarrow S(h,x,h) \in \alpha,$$

and, therefore, the function $S(h,x,h)$ reduces A to α.

From Theorem 1 and Prop. 1, we now obtain the following strengthening of Theorem A.

Theorem A*. *If (α,β) is semi-D.U. and α is r.e. (without β being necessarily r.e.), then α is universal.*

Proof. Suppose (α,β) is semi-D.U. and α is r.e. By our definition of semi-D.U., α and β are disjoint, so $\beta \subseteq \tilde{\alpha}$. Since (α,β) is semi-D.U., so is $(\alpha,\tilde{\alpha})$. Then by Proposition 1, there is a separation function for $(\alpha,\tilde{\alpha})$ and since α is r.e., α is universal by Theorem 1.

§2. Function-Theoretic Analogue.

Next we consider a recursive function-theoretic analogue of Shepherdson's exact separation theorem.

Theorem 2. *If there is a separation function for (α,β) and α and β are both r.e., then (α,β) is D.U.*

Proof. Suppose $S(x,y,z)$ is a separation function for (α,β) and α and β are both r.e. Let (A,B) be any disjoint pair of r.e. sets that we wish to reduce to (α,β). The relation $x \in A \vee S(y,x,y) \in \beta$ and the relation $x \in B \vee S(y,x,y) \in \alpha$ are both r.e., so there is a number h such that for all x and y

1. $((x \in A \vee S(y,x,y) \in \beta) \wedge \sim (x \in B \vee S(y,x,y) \in \alpha)) \Rightarrow$
   $S(h,x,y) \in \alpha$,
2. $((x \in B \vee S(y,x,y) \in \alpha) \wedge \sim (x \in A \vee S(y,x,y) \in \beta)) \Rightarrow$
   $S(h,x,y) \in \beta$.

Taking h for y, it follows by propositional logic that

$$x \in A \leftrightarrow S(h,x,h) \in \alpha$$

and that

$$x \in B \leftrightarrow S(h,x,h) \in \beta,$$

and so the function $S(h,x,h)$ reduces (A,B) to (α,β).

Remark. Theorem 2 provides another sufficient condition for a pair of r.e. sets to be D.U. It, with Proposition 1, of course, has Theorem B as a corollary.

We now see that the Ehrenfeucht-Feferman theorem and the Putnam-Smullyan theorem *can* be proved by Shepherdsonian methods, since we have proved Theorems A and B by Shepherdsonian methods. Our proofs of Theorems 2 and 3 are so close to Shepherdson's proofs of his representation and exact separation theorems that it should be possible to derive both from a common construction. This is, indeed, possible; one way is by using the abstract representation systems of T.F.S. We plan to pursue this further in a companion volume, "Diagonalization and Self-Reference."

We have now given two proofs of Theorems A and B. For our third pair of proofs we will need Kleene's recursion theorem and the author's double analogue (the double recursion theorem), to which we turn in the next two chapters.

§3.   More on the Shepherdson and Putnam-Smullyan Theorems.

A still more direct and simple proof of the Putnam-Smullyan theorem will now be given. It is along *purely* Shepherdson lines and uses virtually no recursion theory (in particular, it avoids even the use of the D.U. pair (U_1, U_2)). Moreover, the proof yields a stronger result—namely that if S is a consistent axiomatizable Rosser system for sets, then for S to be an exact Rosser system for sets, it is enough that for each number h, the function $J(h, x)$ (as a function of x) is strongly definable in S. [And thus it is more than enough that all Σ_0- functions be strongly definable in S.]

We shall first prove some variants of Shepherdson's lemma and theorem that will have other applications later on. The following lemma is a variant of Lemma 2* of Ch. 0. Its proof is similar and left to the reader.

Lemma 2$^\sharp$. *Suppose $R_1(x, y)$ and $R_2(x, y)$ are disjoint relations. Let B_1 be the set of all numbers $J(y, x)$ such that*

$$R_1(x, y) \vee (E_y[\overline{J(y, x)}] \text{ is refutable in } S).$$

Let B_2 be the set of all numbers $J(y, x)$ such that

$$R_2(x, y) \vee (E_y[\overline{J(y, x)}] \text{ is provable in } S).$$

Then if $E_h(v_1)$ is a formula that strongly separates $B_1 - B_2$ from

$B_2 - B_1$ in $\mathcal{S}$, then, for any number n, we have:

(1) $E_h(\overline{J(h,n)})$ is provable in $\mathcal{S}$ iff $R_1(n,h)$.
(2) $E_h(\overline{J(h,n)})$ is refutable in $\mathcal{S}$ iff $R_2(n,h)$.

If, in the above lemma, the relations $R_1(x,y)$ and $R_2(x,y)$ are both r.e., and if $\mathcal{S}$ is axiomatizable, then the sets B_1 and B_2, are r.e., and so by the lemma we get the following variant of Theorem S_2^* of Ch. 0.

Theorem $\mathrm{S}_2^\sharp$. *If $\mathcal{S}$ is a consistent axiomatizable Rosser system for sets, then for any disjoint r.e. relations $R_1(x,y)$ and $\overline{R_2(x,y)}$, there is a formula $E_h(v_1)$ such that for any number n, $E_h(\overline{J(h,n)})$ is provable in $\mathcal{S}$ iff $R_1(n,h)$, and is refutable in $\mathcal{S}$ iff $R_2(n,h)$.*

As a corollary, we have the following variant of Shepherdson's theorem S_2:

Theorem S_2'. *Under the same hypothesis, for any disjoint r.e. sets A_1 and A_2 there is a formula $H(v_1)$ and a number h (in fact the Gödel number of $H(v_1)$) such that for any number n, $H(\overline{J(h,n)})$ is provable iff $n \in A_1$ and is refutable iff $n \in A_2$.*

Proof. By Theorem $\mathrm{S}_2^\sharp$, taking $R_1(x,y)$ iff $x \in A_1$ and $R_2(x,y)$ iff $x \in A_2$.

From Theorem S_2' we then get:

Theorem P.S.*. [1] *Suppose $\mathcal{S}$ is a consistent axiomatizable Rosser system for sets and that for every number h, the function $J(h,x)$ is strongly definable in $\mathcal{S}$. Then $\mathcal{S}$ is an exact Rosser system for sets.*

Proof. Assume hypothesis. Given disjoint r.e. sets A_1 and A_2, take $H(v_1)$ and h as in the conclusion of Theorem S_2'. Since, by hypothesis, the function $J(h,x)$ is definable in $\mathcal{S}$, by Theorem 11.1, Ch. 0, there is a formula $F(v_1)$ such that for every n,

$$F(\overline{n}) \equiv G(\overline{J(h,n)})$$

is provable in $\mathcal{S}$. Then $F(v_1)$ exactly separates (A_1, A_2) in $\mathcal{S}$.

---

[1] An extension of the Putnam-Smullyan theorem.

Chapter VIII

Recursion Theorems

We have proved that the complement of every *completely* productive set (in other words, every generative set) is universal, and this was enough to establish Theorem A of Chapter 6. In Chapter 10 we will prove Myhill's stronger result that the complement of every *productive* set is universal. For this proof, we will need the recursion theorem of this chapter.

Recursion theorems (which can be stated in many forms) have profound applications in recursive function theory and metamathematics, and we shall devote considerable space to their study. To illustrate their rather startling nature, consider the following mathematical "believe-it-or-not's": Which of the following propositions, if true, would surprise you?

1. There is a number n such that $\omega_n = \omega_{n+1}$.
2. There is a number n such that $\omega_n = \omega_{3n^2+4n+7}$.
3. For *any* recursive function $f(x)$, there is a number n such that $\omega_n = \omega_{f(n)}$.
4. There is a number n such that ω_n contains n as its only element.
5. For any recursive function $f(x)$, there is a number n such that $\omega_n = \{f(n)\}$.
6. There is a number h such that for every number x,

$$x \in \omega_h \leftrightarrow h \in \omega_x.$$

[In other words, the set of all x such that ω_x contains h has h as an index!]
7. There is a number h such that $\omega_h = x : R_h(x, h)$.
8. For any r.e. relation $R(x, y)$, there is a number n such that $\omega_n = x : R(x, n)$.

Well, believe it or not, *all* the above propositions are true! At first

94

sight, it might appear that these seemingly coincidental properties are due to some special property of our indexing, but as a matter of fact, they all hold for any indexing we choose, provided the indexing is maximal (in the sense of Exercise 10, Chapter 3).

I.   Weak Recursion Theorems

§1.   The Weak Recursion Theorem.

We now prove statement (8) above—this is the *weak* recursion theorem—and we will see that all the other statements are derivable as corollaries.

Theorem 1—The Weak Recursion Theorem. *For any r.e. relation $R(x, y)$, there is a number n such that $\omega_n = x : R(x, n)$.*

Proof. The relation $R_y(x, y)$ (as a relation between x and y) is r.e., so by the iteration theorem, there is a recursive function $d(y)$ such that for all x and y,

$$x \in \omega_{d(y)} \leftrightarrow R_y(x, y).$$

Now take any r.e. relation $R(x, y)$. Then the relation $R(x, d(y))$ is also r.e.; let m be an index of this relation. Then for all x,

$$x \in \omega_{d(m)} \leftrightarrow R_m(x, m) \leftrightarrow R(x, d(m)).$$

Thus $x \in \omega_{d(m)} \leftrightarrow R(x, d(m))$. And so for all x, $x \in \omega_n \leftrightarrow R(x, n)$, where n is the number $d(m)$.

The above proof revealed more than the statement of the theorem. It will be useful to record this additional information in the following sharper form of Theorem 1.

Theorem 1$^\sharp$. *There is a recursive function $d(y)$ such that for any r.e. relation $R(x, y)$, if m is any index of the relation $R(x, d(y))$, then $\omega_{d(m)} = x : R(x, d(m))$.*

Theorem 1 has the following corollary.
Theorem 1.1—Another form of the weak recursion theorem.

For any recursive function $f(x)$, there is a number n such that

$$\omega_{f(n)} = \omega_n.$$

Proof. Let $R(x, y)$ be the r.e. relation $x \in \omega_{f(y)}$. By Theorem 1 there is a number n such that $\omega_n = x : R(x, n)$. Hence,

$$\omega_n = x : x \in \omega_{f(n)} = \omega_{f(n)}.$$

Exercise 1. Prove the other seven "believe-it-or-not's".

Exercise 2.

1. Without using Theorem 1, prove that for any r.e. relation $R(x_1, \ldots, x_n, y)$, there is a number h such that for all $x_1, \ldots, x_n$,

$$R_h(x_1, \ldots, x_n) \leftrightarrow R(x_1, \ldots, x_n, h).$$

2. Derive (a) as a corollary of Theorem 1 by using the function $J_n(x_1, \ldots, x_n)$.

Exercise 3. Prove the following more general form of Theorem 1.1: For any positive n and any recursive function $f(x)$, there is a number i such that $R_i^n = R_{f(i)}^n$.

§2.   Unsolvable Problems and Rice's Theorem.   The "believe-it-or-not's" mentioned in the last section may seem like somewhat frivolous applications of the recursion theorem. We now turn to a far more serious application.

Let us call a property $P(n)$ *solvable* if the set of all n having the property P is recursive. For example, suppose that $P(n)$ is the property that ω_n is a finite set. To ask whether this property is solvable is to ask whether the set of all indices of all finite r.e. sets is recursive.

Here are some typical properties $P(n)$ that have come up in the literature.

(1) ω_n is infinite.
(2) ω_n is empty.
(3) ω_n is recursive.
(4) ω_n contains the number 5.
(5) $\omega_n = \omega_{17}$.
(6) The relation $R_n(x, y)$ is single-valued (i.e., for all x, there is at most one y such that $R_n(x, y)$).
(7) The relaton $R_n(x, y)$ is functional (for all x, there is exactly one y such that $R_n(x, y)$).

(8) All numbers are in the domain of $R_n(x, y)$.

(9) Infinitely many numbers are in the range of $R_n(x, y)$.

For each of these nine properties, we can ask whether it is solvable; these are typical questions in recursion theory that have been answered by different workers in the field and by individual methods. Rice's theorem, to which we are about to turn, answers a host of such questions in one fell swoop.

Call a number set A *extensional* if A contains, with any index of an r.e. set, all other indices of it as well. Thus, A is extensional iff for every i and j, if $\omega_i = \omega_j$, then $i \in A \leftrightarrow j \in A$. The set N of all natural numbers is obviously extensional and the empty set $\emptyset$ is vacuously extensional; both these sets are recursive. Are there any other recursive extensional sets?

Theorem R—Rice's Theorem.   *The only recursive extensional sets are N and $\emptyset$.*

Lemma. *For any recursive set A other than N or $\emptyset$ there is a recursive function $f(x)$ such that for all x, $x \in A \leftrightarrow f(x) \notin A$.*

Proof of Lemma.   Suppose that A is recursive and that at least one number a is in A, and at least one number b is not in A. Define $f(x)$ as follows:

1. $f(x) = b$, if $x \in A$,
2. $f(x) = a$, if $x \notin A$.

If $x \in A$, then $f(x) = b$; hence $f(x) \notin A$. If $x \notin A$, then $f(x) = a$; hence $f(x) \in A$. So $x \in A \leftrightarrow f(x) \notin A$. The relation $f(x) = y$ is r.e., for it can be written as

$$(x \in A \wedge y = b) \vee (x \in \tilde{A} \wedge y = a)$$

(it is r.e. since A and $\tilde{A}$ are both r.e. sets). Therefore, $f(x)$ is recursive.

Proof of Rice's Theorem.   Let A be any recursive set other than N or $\emptyset$. By the above lemma, there is a recursive function $f(x)$ such that for all x, $x \in A \leftrightarrow f(x) \notin A$.

By Theorem 1.1, there is some number n such that $\omega_n = \omega_{f(n)}$. Also $n \in A \leftrightarrow f(n) \notin A$, so the numbers n and $f(n)$ are indices of the same r.e. set, yet one of them is in A and the other is not. Therefore, A is not extensional.

Applications. Let us consider the first of the nine properties mentioned above—ω_n is infinite. Obviously if ω_i is infinite and $\omega_j = \omega_i$, then ω_j is infinite, so the set of indices of infinite sets is extensional. Also there is at least one i such that ω_i is infinite and at least one i such that ω_i is not infinite. Therefore, the set of indices of all infinite sets satisfies the hypothesis of Rice's theorem and, hence, it is not recursive. The same argument applies to all the other eight properties; hence *none* of them are solvable!

The whole point is that if $\mathcal{C}$ is any non-empty class of r.e. sets which does not contain all r.e. sets, then the set of all indices of all members of $\mathcal{C}$ is not a recursive set.

Exercise 4. Consider the set of all ordered pairs (i,j) such that $\omega_i = \omega_j$. Is this relation recursive?

Exercise 5. Consider the set of all ordered pairs (i,j) such that j is in the range of the relation $R_i(x,y)$. Is this relation recursive?

II.   The Strong Recursion Theorem

§3.   Strong Recursion Theorem.   By Theorem 1, for any i there exists some j such that $\omega_j = x : R_i(x,j)$. Given the number i, can we find j as a recursive function of i? That is, is there a recursive function $\phi(x)$ such that for any i, $\omega_{\phi(i)} = x : R_i(x, \phi(i))$? An affirmative answer (Myhill's fixed point theorem) is a special case of the following theorem.

Theorem 2—The Strong Recursion Theorem. *For any r.e. relation $M(x,y,z)$, there is a recursive function $\phi(y)$ such that for all i, $\omega_{\phi(i)} = x : M(x,i,\phi(i))$.*

We will give two proofs of this important theorem. The first is along traditional lines; the second is quite different and will be a model of subsequent proofs of other recursion theorems.

Proof 1.   We take a recursive function $d(y)$ satisfying Theorem $1^{\sharp}$. Now, given an r.e. relation $M(x,y,z)$, the relation $M(x,z,d(y))$ (as a relation among x, y and z) is r.e. [In λ-notation, we are considering the relation $\lambda x, y, z : M(x,z,d(y))$.] By the iteration theorem, there is a recursive function $t(z)$ such that for all i, the number $t(i)$ is an index of the relation $M(x,i,d(y))$ (i.e. for all x and y, $R_i(x,y) \leftrightarrow M(x,i,d(y))$.) Then by Theorem $1^{\sharp}$ (taking $t(i)$ for m), $\omega_{d(t(i))} = x :$

$M(x, i, d(t(i)))$. And so we take $\phi(y)$ to be the function $d(t(y))$.

Remark. The above proof required two applications of the iteration theorem—one to obtain the recursive function $d(y)$ of Theorem $1^\sharp$ and the other to obtain the function $t(y)$. Our second proof below requires only one application of the iteration theorem. Before turning to this proof, we would like to point out that Theorem 2 can be alternatively derived from Exercise 18, Chapter 3 (whose solution, also involving two applications of the iteration theorem, has been given). By this exercise, there is a recursive function $F(x, y)$ such that for any r.e. relation $M(x, y, z)$, there is a recursive function $t(y)$ such that for all y and z,

$$\omega_{F(t(y),z)} = x : M(x, y, z).$$

Now take any r.e. relation $M(x, y, z)$. The relation $M(x, y, F(z, z))$ is r.e., so there is a recursive function $t(y)$ such that for all y and z,

$$\omega_{F(t(y),z)} = x : M(x, y, F(z, z)).$$

Then

$$\omega_{F(t(y),t(y))} = x : M(x, y, F(t(y), t(y)))$$

and so we take $\phi(y)$ to be $F(t(y), t(y))$.

This is really Proof 1, but the steps are presented in a different order; we, so to speak, diagonalized "at the last minute". The functions $F(y, y)$ and $d(y)$ are actually the same. Also, this proof really used two applications of the iteration theorem (since two were necessary to prove the existence of a master function).

Our next proof of Theorem 2 is more than just "another proof". It uses a device which will be necessary for the proof of Theorem E of the next section. Here is our second proof.

Proof 2. In this proof we use the iteration theorem only once and we get a different "fixed point" function $\phi(y)$ from the one obtained in the first proof.

The relation $R_z(x, y, z)$, as a relation among x, y and z, is r.e., so by the iteration theorem there is a recursive function $t(y, z)$ such that for all y and z,

(1) $\omega_{t(y,z)} = x : R_z(x, y, z)$.

Now take any r.e. relation $M(x, y, z)$. The relation $M(x, y, t(y, z))$ is r.e.; let h be an index. Then for all x, y and z,

$$R_h(x, y, z) \leftrightarrow M(x, y, t(y, z)),$$

and so

$$R_h(x, y, h) \leftrightarrow M(x, y, t(y, h)).$$

By (1), $\omega_{t(y,h)} = x : R_h(x, y, h) = x : M(x, y, t(y, h))$. So

$$\omega_{t(y,h)} = x : M(x, y, t(y, h)).$$

We, therefore, take $\phi(y)$ to be the recursive function $t(y, h)$.

Our second proof reveals a fact which we would like to record on its own right.

Theorem 2$^\sharp$. *There is a recursive function $t(y, z)$ such that for any r.e. relation $M(x, y, z)$, we have for all y,*

$$\omega_{t(y,h)} = x : M(x, y, t(y, h)),$$

where h is any index of the relation $M(x, y, t(y, z))$.

Discussion. Recursion theorems are sometimes referred to as *fixed point* theorems. One usually thinks of fixed points with reference to *functions*; a fixed point of a function $f(x)$ is an element n such that $f(n) = n$. However, we can generalize the notion of a fixed point as follows: Call n a fixed point of a *relation* $R(x, y)$ if $R(n, n)$ holds. [Thus n is a fixed point of a function $f(x)$ iff n is a fixed point of the relation $f(x) = y$.] Now suppose R is an r.e. relation. Define $R^*(m, n)$ to hold if $\omega_m = x : R(x, n)$. What Theorem 1 tells us is that for any r.e. relation $R(x, y)$, there is a fixed point for the relation $R^*(x, y)$. As such, Theorem 1 can be looked at as a relational fixed point theorem.

Now consider a 3-place relation $R(x, y, z)$. Call a function $\phi(x)$ a *fixed point function* for R if for all i, $R(i, \phi(i), \phi(i))$ holds. For any r.e. relation $M(x, y, z)$, define $M^*(i, m, n)$ to hold if $\omega_n = x : M(x, i, m)$. Theorem 2 tells us that for any r.e. relation $M(x, y, z)$, not only is it the case that for every i there is a fixed point for the relation $M^*(i, x, y)$, but also there is a recursive function $\phi(y)$ such that for all i, $M^*(i, \phi(i), \phi(i))$ holds—in other words, there is a recursive fixed point function for the relation $M^*(x, y, z)$.

We called Theorem 1 *weak* because it asserts merely the existence of fixed points. We called Theorem 2 *strong* because it asserts the existence of recursive fixed point functions (at least one for each r.e. relation $M(x, y, z)$).

Theorem 2 has the following corollaries:

Theorem 2.1. *For any r.e. relation $M(x, y, z)$ and any recursive function $g(x)$, there is a recursive function $\phi(y)$ such that for all y,*

$$\omega_{\phi(y)} = x : M(x, y, g(\phi(y))).$$

Proof. By applying Theorem 2 to the r.e. relation $M(x, y, g(z))$.

Theorem 2.2—Myhill's Fixed Point Theorem. *There is a recursive function $\phi(y)$ such that for all y, $\omega_{\varphi(y)} = x : R_y(x, \phi(y))$.*

Proof. By applying Theorem 2 to the r.e. relation $R_y(x, z)$ (as a relation among x, y and z).

Theorem 2.3. *For any recursive function $g(x)$, there is a recursive function $\phi(y)$ such that for all y,*

$$\omega_{\phi(y)} = x : R_y(x, g(\phi(y))).$$

Proof. By applying Theorem 2 to the relation $R_y(x, g(z))$.

Remarks. In Chapter 4 we gave two different proofs that Post's complete set K is generative. We find it of interest that this can also be proved as a corollary of the strong recursion theorem. Let $M(x, y, z)$ be the r.e. relation $J(z, x) \in \omega_y$. Then by Theorem 2, there is a recursive function $\varphi(y)$ such that for all x,

$$x \in \omega_{\varphi(y)} \leftrightarrow J(\varphi(y), x) \in \omega_y.$$

Hence $y \in \omega_{\varphi(y)} \leftrightarrow J(\varphi(y), y) \in \omega_y$. But note $J(\varphi(y), y) \in K \leftrightarrow y \in \omega_{\varphi(y)}$, and so $J(\varphi(y), y) \in K \leftrightarrow J(\varphi(y), y) \in \omega_y$. Thus $J(\varphi(y), y)$ is another generative function for K.

Theorem 2 has the following generalization (which can be derived as a corollary of Theorem 2 or can be proved in an analogous manner).

Theorem 2′. *For any r.e. relation $M(x_1, \ldots, x_n, y_1, \ldots, y_k, z)$ there is a recursive function $\phi(y_1, \ldots, y_k)$ such that for all $x_1, \ldots, x_n$ and $y_1, \ldots, y_k$,*

$$R_{\phi(y_1, \ldots, y_k)}(x_1, \ldots, x_n) \leftrightarrow M(x_1, \ldots, x_n, y_1, \ldots, y_k, \phi(y_1, \ldots, y_k)).$$

Proof. See Exercises 6, 7 and 8 below.

Exercise 6. Prove Theorem 2′ along the following lines: Given $n > 0$, there is a recursive function $d(z)$ such that for all $x_1 \ldots, x_n$, $R_{d(z)}(x_1, \ldots, x_n) \leftrightarrow R_z(x_1, \ldots, x_n, z)$. Then, given

$$M(x_1, \ldots, x_n, y_1, \ldots, y_k, z),$$

take a recursive function $t(y_1, \ldots, y_k)$ such that for all x_1, x_n, y_1, y_k and z,

$$R_{t(y_1,\ldots,y_k)}(x_1, \ldots, x_n) \leftrightarrow M(x_1, \ldots, x_n, y_1, \ldots, y_k, d(z)).$$

Take $\phi(y_1, \ldots, y_n)$ to be $dt(y_1, \ldots, y_n)$ and show that the function ϕ works.

Exercise 7. Alternatively, take a recursive function $H(y_1, \ldots, y_k, z)$ such that for all $x_1, \ldots, x_n, y_1, \ldots, y_k$ and z,

$$R_{H(y_1,\ldots,y_k,z)}(x_1, \ldots, x_n) \leftrightarrow R_z(x_1, \ldots, x_n, y_1, \ldots, y_k, z).$$

Let h be an index of the relation

$$M(x_1, \ldots, x_n, y_1, \ldots, y_k, H(y_1, \ldots, y_k, z))$$

and show that the function $H(y_1, \ldots y_k, h)$ works.

Exercise 8. Alternatively, given an r.e. relation

$$M(x_1, \ldots, x_n, y_1, \ldots, y_k, z),$$

show there is an r.e. relation $R(x, y, z)$ such that for all $x_1, \ldots, x_n$, $y_1, \ldots, y_k$ and z,

$$R(J_n(x_1, \ldots, x_n), J_k(y_1, \ldots, y_k), z) \leftrightarrow M(x_1, \ldots, x_n, y_1, \ldots, y_k, z).$$

Applying Theorem 2 to the relation $R(x, y, z)$ there is a recursive function $\psi(y)$ such that for all y and $\omega_{\psi(y)} = x$, $R(x, y, \phi(y))$). Take $\phi(y_1, \ldots, y_k)$ to be $\psi J(y_1, \ldots, y_k)$ and show that this function works.

Exercise 9. Using Theorem 2, prove that for any r.e. relation $R(x, y)$ and any r.e. set A, there is a recursive function $\phi(y)$ such that

1. For all i in A, $\omega_{\phi(i)} = x : R(x, \phi(i))$.
2. For all $i \notin A$, $\omega_{\phi(i)} = \emptyset$.

Exercise 10—Myhill. Using the exercise above, show that for any recursive function $f(x)$ and any r.e. set A, there is a recursive function $\phi(y)$ such that for any y,

1. If $y \in A$, then $\omega_{\phi(y)} = \{f(\phi(y))\}$.
2. If $y \notin A$, then $\omega_{\phi(y)} = \emptyset$.

Exercise 11. Prove that for any recursive function $g(y)$, there is a recursive function $h(y)$ such that for all y, $\omega_{h(y)} = \omega_y \cap \{g(h(y))\}$.

Exercise 12. Show that for any recursive function $G(x, y)$, there is a recursive function $\phi(x)$ such that for all x, $\omega_{G(x,\phi(x))} = \omega_{\phi(x)}$.

III.   An Extended Recursion Theorem

§4.   We recently thought of an extension of Theorem 2 (Theorem E below) that bears an interesting relation to the double recursion theorems to be considered in the next chapter. It might superficially appear that this theorem can be obtained as a corollary of Theorem 2, but if it can, we do not see how.

We first consider a question. Given an r.e. relation $M(x, y, z)$ and a recursive function $g(x)$, we know by Theorem 2.1 that there is a recursive function $\phi(y)$ such that $\omega_{\phi(y)} = x : M(x, y, g(\phi(y)))$. Is there necessarily a function $\phi(y)$ such that $\omega_{\phi(y)} = x : M(x, y, \phi(g(y)))$? We see no way to prove this using Theorem 2, but we *can* prove it by a simple modification of our second proof of Theorem 2. More generally, we will prove:

Theorem E—The Extended Recursion Theorem.   *For any r.e. relation $M(x, y, z_1, \ldots, z_n)$ and any n-tuple of recursive functions $g_1(y), \ldots, g_n(y)$, there is a recursive function $\phi(y)$ such that for all y and $z_1, \ldots, z_n$,*

$$\omega_{\phi(y)} = x : M(x, y, \phi(g_1(y)), \ldots, \phi(g_n(y))).$$

Proof. As with the second proof of Theorem 2, we take a recursive function $t(y, z)$ such that $\omega_{t(y,z)} = x : R_z(x, y, z)$. We take an index h of the r.e. relation $M(x, y, t(g_1(y), z), \ldots, t(g_n(y), z))$ and so $\omega_{t(y,h)} = x : M(x, y, t(g_1(y), h), \ldots, t(g_n(y), h))$. And so we take $\phi(y) = t(y, h)$.

We note that Theorem 2 is the special case of Theorem E for $n = 1$ and $g_1(y) = y$. For $n = 2$ and $g_1(y)$ the identity function, we have

Corollary. *For any r.e. relation $M(x, y, z_1, z_2)$ and any recursive function $g(y)$, there is a recursive function $\phi(y)$ such that for all y,*

$$\omega_{\phi(y)} = x : M(x, y, \phi(y), \phi(g(y))).$$

Discussion. Our proof of Theorem E is hardly different from our second proof of Theorem 2. Concerning our two different proofs of Theorem 2, our first proof utilized the fact that composition of recursive functions is recursive, which our second proof did not! On the other hand, our second proof utilized the fact that for any recursive function $f(x, y)$ and any number n, the function $f(x, n)$ (as a function of x) is recursive, whereas our first proof did not. In a more abstract setting, these two proofs give rise to distinct theorems (a topic which we will pursue in a companion volume *Diagonalization and Self-Reference*).

Chapter IX

Symmetric and Double Recursion Theorems

I. Double Recursion Theorems

We have proved that every *completely* E.I. pair of r.e. sets is D.U. In the next chapter we will show the stronger result that every E.I. pair of r.e. sets is D.U. The proof of this is based on the *double recursion theorem* of this chapter.

Our original formulation of the double recursion theorem (T.F.S.) required the recursive pairing function $J(x, y)$, not only for its proof, but for its very statement. In this chapter, we give an improved version whose statement and proof are independent of J, K and L.[1] In Part III of this chapter, we compare our new version with the original J, K and L version and show that they are easily interderivable.

§1. The Weak Double Recursion Theorem.

Consider two r.e. relations, $M_1(x, y, z)$ and $M_2(x, y, z)$. For any number b, the relation $M_1(x, y, b)$ is an r.e. relation between x and y, and so by the weak recursion theorem (Theorem 1, Chapter 8), there is a number a such that $\omega_a = x : M_1(x, a, b)$. Likewise, for any number a, the relation $M_2(x, a, y)$ is an r.e. relation between x and y, and so there is a number b such that $\omega_b = x : M_2(x, a, b)$. Our next theorem tells us that we can choose a and b so that both these conditions hold simultaneously.

Theorem 1—The Weak Double Recursion Theorem. *For any r.e. relations $M_1(x, y, z)$ and $M_2(x, y, z)$, there are numbers a*

---

[1] In applications to the theory of double productivity and effective inseparability, the use of recursive pairing functions is actually a quite unnecessary detour.

and b such that

(1) $\omega_a = x : M_1(x, a, b)$,
(2) $\omega_b = x : M_2(x, a, b)$.

In preparation for the strong double recursion theorem of the next section, we will state and prove Theorem 1 in the following sharper form.

Theorem 1$^\sharp$. *There is a recursive function $t(y, z)$ such that for any r.e. relations $M_1(x, y, z)$ and $M_2(x, y, z)$, if c is any index of the relation $M_1(x, t(y, z), t(z, y))$ and d is any index of the relation $M_2(x, t(z, y), t(y, z))$, then,*

(1) $\omega_{t(c,d)} = x : M_1(x, t(c, d), t(d, c))$,
(2) $\omega_{t(d,c)} = x : M_2(x, t(c, d), t(d, c))$.

Proof. By the iteration theorem, there is a recursive function $t(y, z)$ such that for all y and z, $\omega_{t(y,z)} = x : R_y(x, y, z)$.

Now take any two r.e. relations $M_1(x, y, z)$ and $M_2(x, y, z)$. Let c be an index of the r.e. relation $M_1(x, t(y, z), t(z, y))$ and let d be an index of the relation $M_2(x, t(z, y), t(y, z))$. Then

1. $\omega_{t(c,d)} = x : R_c(x, c, d) = x : M_1(x, t(c, d), t(d, c))$,
2. $\omega_{t(d,c)} = x : R_d(x, d, c) = x : M_2(x, t(c, d), t(d, c))$.

Of course, Theorem 1 follows from Theorem 1$^\sharp$ by taking

$$a = t(c, d) \text{ and } b = t(d, c).$$

As a corollary of Theorem 1, we have

Theorem 1.1. *For any two r.e. relations $R_1(x, y)$ and $R_2(x, y)$ and any recursive function $g(x, y)$, there are numbers a and b such that*

(1) $\omega_a = x : R_1(x, g(a, b))$.
(2) $\omega_b = x : R_2(x, g(a, b))$.

Proof. By applying Theorem 1 to the r.e. relations $R_1(x, g(y, z))$ and $R_2(g(y, z))$.

As another corollary we have

Theorem 1.2—A weak Double Myhill Theorem. *For any recursive functions $g_1(x, y), g_2(x, y)$ there are numbers a and b such that,*

$$\omega_a = \omega_{g_1(a,b)} \text{ and } \omega_b = \omega_{g_2(a,b)}.$$

Proof. Exercise.

§2.   The Strong Double Recursion Theorem.   We next
show that in Theorem 1 the numbers a and b can be found as recursive functions of the indices i and j of the relations $M_1(x, y, z)$ and $M_2(x, y, z)$ (Theorem 2.2 below). This will be seen to be a consequence of the following more general theorem.

Theorem 2—The Strong Double Recursion Theorem. *For any r.e. relations $M_1(x, y_1, y_2, z_1, z_2)$ and $M_2(x, y_1, y_2, z_1, z_2)$, there are recursive functions $t_1(y_1, y_2)$ and $t_2(y_1, y_2)$ such that for all y_1 and y_2,*

(1) $\omega_{t_1(y_1,y_2)} = x : M_1(x, y_1, y_2, t_1(y_1, y_2), t_2(y_1, y_2))$,
(2) $\omega_{t_2(y_1,y_2)} = x : M_2(x, y_1, y_2, t_1(y_1, y_2), t_2(y_1, y_2))$.

Proof. We take a recursive function $t(y, z)$ satisfying Theorem $1^{\sharp}$. Now consider any r.e. relations

$$M_1(x, y_1, y_2, z_1, z_2) \text{ and } M_2(x, y_1, y_2, z_1, z_2).$$

By the iteration theorem, there are recursive functions $\phi_1(y_1, y_2)$ and $\phi_2(y_1, y_2)$ such that for any numbers i and j, $\phi_1(i, j)$ is an index of the r.e. relation $M_1(x, i, j, t(y, z), t(z, y))$ (as a relation between x, y and z) and $\phi_2(i, j)$ is an index of the relation $M_2(x, i, j, t(z, y), t(y, z))$. Then by Theorem $1^{\sharp}$ we have (1) and (2) above, taking $t_1(y_1, y_2)$ to be $t(\phi_1(y_1, y_2), \phi_2(y_1, y_2))$ and $t_2(y_1, y_2)$ to be $t(\phi_2(y_1, y_2), \phi_1(y_1, y_2))$.

Remark. The proof above used three applications of the iteration theorem; one to obtain the function $t(y, z)$ of Theorem $1^{\sharp}$ and the other two to obtain the functions $\phi_1(y_1, y_2)$ and $\phi_2(y_1, y_2)$. We will later give an alternative proof which requires only one application of the iteration theorem.

Theorem 2.1. *For any r.e. relations*

$$M_1(x, y, z_1, z_2) \text{ and } M_2(x, y, z_1, z_2),$$

there are recursive functions $t_1(y_1, y_2)$ and $t_2(y_1, y_2)$ such that for all y_1 and y_2,

(1) $\omega_{t_1(y_1,y_2)} = x : M_1(x, y_1, t_1(y_1, y_2), t_2(y_1, y_2))$,
(2) $\omega_{t_2(y_1,y_2)} = x : M_2(x, y_2, t_1(y_1, y_2), t_2(y_1, y_2))$.

Proof. Given the r.e. relations $M_1(x, y, z_1, z_2)$ and $M_2(x, y, z_1, z_2)$, define two new relations by: $M_1'(x, y_1, y_2, z_1, z_2)$ iff $M_1(x, y_1, z_1, z_2)$

and $M_2'(x, y_1, y_2, z_1, z_2)$ iff $M_2(x, y_2, z_1, z_2)$. Then apply Theorem 2 to the relations M_1' and M_2'.

Theorem 2.2—A Double Analogue of Myhill's Fixed-Point Theorem. There are recursive functions $t_1(y_1, y_2)$ and $t_2(y_1, y_2)$ such that for all y_1 and y_2,

(1) $\omega_{t_1(y_1, y_2)} = x : R_{y_1}(x, t_1(y_1, y_2), t_2(y_1, y_2))$,
(2) $\omega_{t_2(y_1, y_2)} = x : R_{y_2}(x, t_1(y_1, y_2), t_2(y_1, y_2))$.

Proof. By Theorem 2.1, taking $M_1(x, y, z_1, z_2)$ and $M_2(x, y, z_1, z_2)$ to both be the r.e. relation $R_y(x, z_1, z_2)$.

Theorem 2.3. *For r.e. relations $M_1(x, y, z_1, z_2)$ and $M_2(x, y, z_1, z_2)$ there are recursive functions $\phi_1(y)$ and $\phi_2(y)$ such that for all y,*

(1) $\omega_{\phi_1(y)} = x : M_1(x, y, \phi_1(y), \phi_2(y))$.
(2) $\omega_{\phi_2(y)} = x : M_2(x, y, \phi_1(y), \phi_2(y))$.

Proof. By Theorem 2.1, taking $\phi_1(y) = t_1(y, y)$ and $\phi_2(y) = t_2(y, y)$

Theorem 2.4. *For any two r.e. relations $M_1(x, y, z)$ and $M_2(x, y, z)$ and any recursive function $g(x, y)$, there are recursive functions $\phi_1(y)$ and $\phi_2(y)$ such that for all y,*

(1) $\omega_{\phi_1(y)} = x : M_1(x, y, g(\phi_1(y), \phi_2(y)))$,
(2) $\omega_{\phi_2(y)} = x : M_2(x, y, g(\phi_1(y), \phi_2(y)))$.

Proof. By Theorem 2.3 applied to the r.e. relations $M_1(x, y, g(z_1, z_2))$ and $M_2(x, y, g(z_1, z_2))$.

Theorem 2.5. *For r.e. relations $M_1(x, y, z)$ and $M_2(x, y, z)$ and recursive function $g(x, y)$, there are recursive functions $t_1(y_1, y_2)$ and $t_2(y_1, y_2)$ such that for all y_1 and y_2,*

(1) $\omega_{t_1(y_1, y_2)} = x : M_1(x, y_1, g(t_1(y_1, y_2), t_2(y_1, y_2)))$,
(2) $\omega_{t_2(y_1, y_2)} = x : M_2(x, y_2, g(t_1(y_1, y_2), t_2(y_1, y_2)))$.

Proof. By applying 2.1 to the r.e. relations

$$M_1(x, y, g(z_1, z_2)) \text{ and } M_2(x, y, g(z_1, z_2))$$

From Theorem 2.5 (taking both $M_1(x, y, z)$ and $M_2(x, y, z)$ to be the relation $R_y(x, z)$), we have,

Theorem 2.6. *For recursive function $g(x, y)$, there are recursive functions $t_1(y_1, y_2)$ and $t_2(y_1, y_2)$ such that,*

(1) $\omega_{t_1(y_1,y_2)} = x : R_{y_1}(x, g(t_1(y_1, y_2), t_2(y_1, y_2)))$,
(2) $\omega_{t_2(y_1,y_2)} = x : R_{y_2}(x, g(t_1(y_1, y_2), t_2(y_1, y_2)))$.

Exercise 1. Why is the weak double recursion theorem an immediate corollary of the strong double recursion theorem?

Exercise 2. Prove the following (weak) "triple" recursion theorem: For r.e. relations $M_1(x, y, z, w)$, $M_2(x, y, z, w)$ and $M_3(x, y, z, w)$, there are numbers a, b and c such that,

  1. $\omega_a = x : M_1(x, a, b, c)$,
  2. $\omega_b = x : M_2(x, a, b, c)$,
  3. $\omega_c = x : M_3(x, a, b, c)$.

II. A Symmetric Recursion Theorem

§3. We recently hit on the following theorem whose proof makes use of only one application of the iteration theorem and which yields the strong double recursion theorem as a corollary.

Theorem S—The Symmetric Recursion Theorem. *For r.e. relation $M(x, z, y_1, y_2, w_1, w_2)$, there is a corresponding recursive function $t(z_1, z_2, y_1, y_2)$ (which we might call a symmetric function for M) such that for all z_1, z_2, y_1 and y_2,*

$$\omega_{t(z_1,z_2,y_1,y_2)} = x : M(x, z_1, y_1, y_2, t(z_1, z_2, y_1, y_2), t(z_2, z_1, y_1, y_2)).$$

Proof. By the iteration theorem, there is a recursive function

$$g(z_1, z_2, y_1, y_2, w)$$

such that for all z_1, z_2, y_1, y_2 and w,

$$\omega_{g(z_1,z_2,y_1,y_2,w)} = x : R_w(x, z_1, z_2, y_1, y_2, w).$$

Now let h be an index of the r.e. relation

$$M(x, z_1, y_1, y_2, g(z_1, z_2, y_1, y_2, w), g(z_2, z_1, y_1, y_2, w)).$$

Then take

$$t(z_1, z_2, y_1, y_2) = g(z_1, z_2, y_1, y_2, h).$$

    Applying Theorem S to the r.e. relation $R_z(x, y_1, y_2, w_1, w_2)$, we get

Theorem SM–A Symmetric Form of Myhill's Fixed-Point Theorem. There is a recursive function $S(z_1, z_2, y_1, y_2)$ such that,

$$\omega_{S(z_1,z_2,y_1,y_2)} = x : R_{z_1}(x, y_1, y_2, S(z_1, z_2, y_1, y_2), S(z_2, z_1, y_1, y_2)).$$

Such a function $S(z_1, z_2, y_1, y_2)$ we will call a *symmetric Myhill function*. The existence of such a function immediately yields the following variant of Theorem 2.

Theorem 2°. *For any two r.e. relations M_1 and M_2 of five arguments, there are recursive functions $t_1(y_1, y_2)$ and $t_2(y_1, y_2)$ such that,*

(1) $\omega_{t_1(y_1,y_2)} = x : M_1(x, y_1, y_2, t_1(y_1, y_2), t_2(y_1, y_2)),$
(2) $\omega_{t_2(y_1,y_2)} = x : M_2(x, y_1, y_2, t_2(y_1, y_2), t_1(y_1, y_2)).$

Proof. Let a_1 and a_2 be respective indices of M_1 and M_2. Then take

$$t_1(y_1, y_2) = S(a_1, a_2, y_1, y_2)$$

and

$$t_2(y_1, y_2) = S(a_2, a_1, y_1, y_2),$$

where S is a symmetric Myhill function.

Of course, Theorem 2 is easily obtainable from Theorem 2°: Given r.e. relations $M_1(x, y_1, y_2, w_1, w_2)$ and $M_2(x, y_1, y_2, w_1, w_2)$, apply Theorem 2° to the relations M_1 and $M_2{}'$, where

$$M_2{}'(x, y_1, y_2, w_1, w_2)$$

is the relation $M_2(x, y_1, y_2, w_2, w_1)$.

Another theorem that yields Theorem 2 more directly (it avoids having to go via Theorem 2°) is Theorem N of §8. The reader can turn to this directly, if desired.

III. Double Recursion with a Pairing Function

Our original formulation (1961) of the double recursion theorem was in terms of the recursive pairing function $J(x, y)$ and its inverse functions Kx and Lx. Here is the original version.

Theorem D. *For any two r.e. relations $M_1(x, y, z)$ and $M_2(x, y, z)$, there is a recursive function $\phi(y)$ such that for all y,*

(1) $\omega_{K\phi y} = x : M_1(x, y, \phi(y)),$

(2) $\omega_{L\phi y} = x : M_2(x, y, \phi(y))$.

We will give an improved version of our original proof of Theorem D which brings to light an interesting feature in its own right (Theorem 3 below).

§4. Double Master Functions.

By a *double master function*, we shall mean a recursive function $G(y, z)$ such that for any r.e. relations $M_1(x, y, z)$ and $M_2(x, y, z)$ there is a recursive function $t(y)$ such that for all y and z, the following two conditions hold:

(1) $\omega_{KG(ty,z)} = x : M_1(x, y, z)$,
(2) $\omega_{LG(ty,z)} = x : M_2(x, y, z)$.

Theorem 3. *There exists a double master function.*

Proof. By Theorem 2.1, Ch. 3, there exists a master function $F(x, y)$. We let $G(y, z)$ be the recursive function $J(F(Ky, z), F(Ly, z))$. Then $KG(y, z) = F(Ky, z)$ and $LG(y, z) = F(Ly, z)$. We show that $G(y, z)$ is a double master function.

Consider any two r.e. relations $M_1(x, y, z)$ and $M_2(x, y, z)$. Since $F(y, z)$ is a master function, there are recursive functions $t_1(y)$ and $t_2(y)$ such that $x : M_1(x, y, z) = \omega_{F(t_1 y, z)}$ and $x : M_2(x, y, z) = \omega_{F(t_2 y, z)}$. We let $t(y) = J(t_1(y), t_2(y))$, and so we have $t_1 y = Kty$ and $t_2 y = Lty$. Therefore,

(1) $x : M_1(x, y, z) = \omega_{F(Kty,z)} = \omega_{KG(ty,z)}$,
(2) $x : M_2(x, y, z) = \omega_{F(Lty,z)} = \omega_{LG(ty,z)}$.

This concludes the proof.

Now that we have a double master function $G(y, z)$, the proof of Theorem D is easy. Given two r.e. relations $M_1(x, y, z)$ and $M_2(x, y, z)$, the relations $M_1(x, y, G(z, z))$ and $M_2(x, y, G(z, z))$ are r.e.; hence there is a recursive function $t(y)$ such that,

(1) $\omega_{KG(ty,z)} = x : M_1(x, y, G(z, z))$,
(2) $\omega_{LG(ty,z)} = x : M_2(x, y, G(z, z))$.

In (1) and (2), we replace z by $t(y)$ and we have,

(1)' $\omega_{KG(ty,ty)} = x : M_1(x, y, G(ty, ty))$.
(2)' $\omega_{LG(ty,ty)} = x : M_2(x, y, G(ty, ty))$.

And so we take $\phi(y) = G(t(y), t(y))$.

Exercise 3. The above proof sort of "doubled up" on our first proof of the (strong) recursion theorem (Theorem 2 of Ch. 8). We can also "double up" on the second proof of that theorem and obtain Theorem D in the following manner: Take recursive functions $t_1(y, z)$ and $t_2(y, z)$ such that $\omega_{t_1(y,z)} = x : R_{Kz}(x, y, z)$ and $\omega_{t_2(y,z)} = x : R_{Lz}(x, y, z)$. Then take $t(y, z) = J(t_1(y, z), t_2(y, z))$. Let h_1 and h_2 be respective indices of the relations $M_1(x, y, t(y, z))$ and $M_2(x, y, t(y, z))$ and let $h = J(h_1, h_2)$. Now take $\phi(y)$ to be $t(y, h)$ and show that the function $\phi(y)$ works.

§5. Theorem D and Theorem 2 Compared.

From Theorem D, we can obtain an alternative proof of the strong double recursion theorem (Theorem 2). Given two r.e. relations

$$M_1(x, y_1, y_2, z_1, z_2) \text{ and } M_2(x, y_1, y_2, z_1, z_2),$$

apply Theorem D to the two relations

$$M_1(x, Ky, Ly, Kz, Lz) \text{ and } M_2(x, Ky, Ly, Kz, Lz)$$

(as relations among x, y and z) to get a recursive function $\phi(y)$ such that for all y,

(1) $\omega_{K\phi y} = x : M_1(x, Ky, Ly, K\phi y, L\phi y)$,
(2) $\omega_{L\phi y} = x : M_2(x, Ky, Ly, K\phi y, L\phi y)$.

Then take $t_1(y_1, y_2)$ to be $K\phi J(y_1, y_2)$, and also take $t_2(y_1, y_2)$ to be $L\phi J(y_1, y_2)$. The reader can easily verify that the functions $t_1(y_1, y_2)$ and $t_2(y_1, y_2)$ work.

It is also possible to proceed in the reverse direction and derive Theorem D as a corollary of Theorem 2. Here is the method.

Theorem 2 has Theorem 2.4 as a corollary, so we will derive Theorem D from Theorem 2.4 directly. We take $g(x, y)$ to be $J(x, y)$ and so there are recursive functions $\phi_1(y)$ and $\phi_2(y)$ such that for all y,

(1) $\omega_{\phi_1(y)} = x : M_1(x, y, J(\phi_1(y), \phi_2(y)))$,
(2) $\omega_{\phi_2(y)} = x : M_2(x, y, J(\phi_1(y), \phi_2(y)))$.

We take $\phi(y)$ to be $J(\phi_1(y), \phi_2(y))$, and so $\phi_1(y) = K\phi y$ and $\phi_2(y) = L\phi y$, and we have,

(1)$'$ $\omega_{K\phi y} = x : M_1(x, y, \phi(y))$,
(2)$'$ $\omega_{L\phi y} = x : M_2(x, y, \phi(y))$.

Exercise 4. Using the double recursion theorem, prove that for any two r.e. sets A_1 and A_2 and any two r.e. relations $R_1(x,y)$ and $R_2(x,y)$ and any recursive function $g(x,y)$, there are recursive functions $\phi_1(y)$ and $\phi_2(y)$ such that for all y,

1. $y \in A_1 \Rightarrow \omega_{\phi_1(y)} = x : R_1(x, g(\phi_1(y), \phi_2(y)))$,
2. $y \notin A_1 \rightarrow \omega_{\phi_1(y)} = \emptyset$,
3. $y \in A_2 \Rightarrow \omega_{\phi_2(y)} = x : R_2(x, g(\phi_1(y), \phi_2(y)))$,
4. $y \notin A_2 \Rightarrow \omega_{\phi_1(y)} = \emptyset$.

Exercise 5. Show that for any r.e. sets α, β, A and B, and any recursive function $g(x,y)$, there are recursive functions $\phi_1(y)$ and $\phi_2(y)$ such that for all y,

1. $y \in B \Rightarrow \omega_{\phi_1(y)} = \alpha \cup \{g(\phi_1(y), \phi_2(y))\}$,
2. $y \notin B \Rightarrow \omega_{\phi_1(y)} = \alpha$,
3. $y \in A \Rightarrow \omega_{\phi_2(y)} = \beta \cup \{g(\phi_1(y), \phi_2(y))\}$,
4. $y \notin A \Rightarrow \omega_{\phi_2(y)} = \beta$.

Exercise 6. Show that for any recursive function $g(x,y)$, there are recursive functions $t_1(y_1, y_2)$ and $t_2(y_1, y_2)$ such that for all y_1 and y_2,

1. $\omega_{t_1(y_1,y_2)} = \omega_{y_1} \cap \{g(t_1(y_1, y_2), t_2(y_1, y_2))\}$,
2. $\omega_{t_2(y_1,y_2)} = \omega_{y_2} \cap \{g(t_1(y_1, y_2), t_2(y_1, y_2))\}$.

IV.   Further Topics

§6.   Applications of the Extended Recursion Theorem.

It is possible to derive the strong double recursion theorem from the strong recursion theorem (two applications are required—cf. §7 of this chapter). We do not know, however, whether the symmetric recursion theorem (which appears to be stronger than the strong double recursion theorem) can be derived as a corollary of the strong recursion theorem. But it *can* be derived from the extended recursion theorem of the last chapter, as we will now see. Along the way we will derive a variant of Theorem S which bears much the same relationship to Theorem S as Theorem 2 bears to Theorem D.

The Bar Symmetric Recursion Theorem.   For any number x, we define $\bar{x}$ to be $J(Lx, Kx)$. Thus for any numbers x_1 and x_2, we have $\overline{J(x_1, x_2)} = J(x_2, x_1)$. Obviously $\bar{\bar{x}} = x$ and $K\bar{x} = Lx$ and

$L\overline{x} = Kx.$

Theorem B—The Bar Symmetric Recursion Theorem.

For any r.e. relation $M(x, z, y, w_1, w_2)$, there is a recursive function $\psi(z, y)$ such that for all z and y,

$$\omega_{\psi(z,y)} = x : M(x, z, y, \psi(z, y), \psi(\overline{z}, y)).$$

Proof. Define $g(x) = J(\overline{Kx}, Lx)$. Then for any y and z,

$$gJ(z, y) = J(\overline{z}, y).$$

Of course the function g is recursive.

Now, given an r.e. relation $M(x, z, y, w_1, w_2)$, let $M_1(x, y, z_1, z_2)$ be the r.e. relation $M(x, Ky, Ly, z_1, z_2)$. Applying the corollary of Theorem E (last chapter) to M_1, we have a recursive function $\phi(w)$ such that,

$$\omega_{\phi(w)} = x : M_1(x, w, \phi(w), \phi g(w)).$$

We take $J(z, y)$ for w and we have,

$$
\begin{aligned}
\omega_{\phi J(z,y)} &= x : M_1(x, J(z, y), \phi J(z, y), \phi g J(z, y)) \\
&= x : M_1(x, J(z, y), \phi J(z, y), \phi J(\overline{z}, y)) \\
&= x : M(x, z, y, \phi J(z, y), \phi J(\overline{z}, y))
\end{aligned}
$$

And so we take $\psi(z, y) = \phi J(z, y)$.

From Theorem B we can obtain an alternative proof of Theorem S as follows: Given an r.e. relation $M(x, z, y_1, y_2, w_1, w_2)$, we let $M_1(x, z, y, w_1, w_2)$ be the r.e. relation $M(x, Kz, Ky, Ly, w_1, w_2)$. Thus $M_1(x, J(z_1, z_2), J(y_1, y_2), w_1, w_2)$ is equivalent to

$$M(x, z_1, y_1, y_2, w_1, w_2).$$

We then apply Theorem B to the relation M_1 and get a recursive function $\psi(z, y)$ such that $\omega_{\psi(z,y)} = x : M_1(x, z, y, \psi(z, y), \psi(\overline{z}, y))$. We then take $t(z_1, z_2, y_1, y_2) = \psi(J(z_1, z_2), J(y_1, y_2))$, and the reader can easily verify that this function t works (the crucial point being that

$$t(z_2, z_1, y_1, y_2) = \psi(\overline{J(z_1, z_2)}, J(y_1, y_2)),$$

since

$$\overline{J(z_1, z_2)} = J(z_2, z_1)).$$

Remarks. We got from the extended recursion theorem to the symmetric recursion theorem via Theorem B. One can get directly from Theorem E to Theorem S as follows: We take the recursive "quadrupling" function $J_4(x_1, x_2, x_3, x_4)$ (cf. §5, Chapter 3) and its inverse functions $K_1^4, \ldots, K_4^4$—which we will now write $\sigma_1, \ldots, \sigma_4$. We let $g(x) = J_4(\sigma_2 x, \sigma_1 x, \sigma_3 x, \sigma_4 x)$ and, thus, for any numbers $x_1, \ldots, x_4$,

$$gJ_4(x_1, x_2, x_3, x_4) = J_4(x_1, x_2, x_4, x_3).$$

Then given an r.e. relation $M(x, z, y_1, y_2, w_1, w_2)$, we let the relation $M_1(x, y, w_1, w_2)$ be the relation $M(x, \sigma_1 y, \sigma_3 y, \sigma_4 y, w_1, w_2)$ and apply corollary of Theorem E to M_1, thus getting a recursive function $\phi(y)$ such that $\omega_{\phi(y)} = x : M_1(x, y, \phi y, \phi g y)$. We then take

$$t(z_1, z_2, y_1, y_2) = \phi J_4(z_1, z_2, y_1, y_2),$$

and the reader can verify that this function works.

§7. The (Strong) Single and Double Recursion Theorems Compared.

We have seen in the literature derivations of the weak double recursion theorem from the strong (single) recursion theorem (cf., e.g. Rogers [1967]). It is also possible to obtain the strong double recursion theorem from the strong recursion theorem as follows: We use the strong recursion theorem in the more general form of Theorem 2′, Ch. 8. We need two applications of it—one for $n = 1, k = 3$, and one for $n = 1, k = 2$—i.e., we need the following special cases.

(A) For any r.e. relation $M(x, y_1, y_2, y_3, z)$, there is a recursive function $\phi(y_1, y_2, y_3)$ such that for all y_1, y_2 and y_3,

$$\omega_{\phi(y_1, y_2, y_3)} = x : M(x, y_1, y_2, y_3, \phi(y_1, y_2, y_3)).$$

(B) For any r.e. relation $M(x, y_1, y_2, z)$, there is a recursive function $t(y_1, y_2)$ such that for all y_1 and y_2,

$$\omega_{t(y_1, y_2)} = x : M(x, y_1, y_2, t(y_1, y_2)).$$

Now to derive the strong double recursion theorem. Let

$$M_1(x, y_1, y_2, z_1, z_2) \text{ and } M_2(x, y_1, y_2, z_1, z_2)$$

be two r.e. relations. Applying (A) to the relation M_2, there is a recursive function $\phi(y_1, y_2, y_3)$ such that for all y_1, y_2 and y_3,

(1) $\omega_{\phi(y_1, y_2, y_3)} = x : M_2(x, y_1, y_2, y_3, \phi(y_1, y_2, y_3)).$

Next we apply (B) to the r.e. relation $M_1(x, y_1, y_2, z, \phi(y_1, y_2, z))$ (as a relation among x, y_1, y_2 and z), and obtain a recursive function $t_1(y_1, y_2)$ such that for all y_1 and y_2,

(2) $\omega_{t_1(y_1,y_2)} = x : M_1(x, y_1, y_2, t_1(y_1, y_2), \phi(y_1, y_2, t_1(y_1, y_2)))$.

Then, in (1), we substitute $t_1(y_1, y_2)$ for y_3 and get,

(1)$'$ $\omega_{\phi(y_1,y_2,t_1(y_1,y_2))} = x : M_2(x, y_1, y_2, t_1(y_1, y_2), \phi(y_1, y_2, t_1(y_1, y_2)))$.

We, therefore, take $t_2(y_1, y_2)$ to be $\phi(y_1, y_2, t_1(y_1, y_2))$ and we see that the pair $(t_1(y_1, y_2), t_2(y_1, y_2))$ works.

We admire the above proof for its ingenuity, but we find the direct proof of the double recursion theorem, or the approach using the symmetric recursion theorem, to be more straightforward and intuitive. Better yet, we have Theorem N of the next section.

§8. A Very Nice Function.

In deriving Theorem 2 from the symmetric recursion theorem, we first had to prove Theorem 2° and then make a "switch". The following theorem yields Theorem 2 more elegantly.

Theorem N. *There is a recursive function* $N(z, z_1, z_2, y_1, y_2)$ *such that for all* z, z_1, z_2, y_1 *and* y_2,

$$\omega_{N(z,z_1,z_2,y_1,y_2)} = \\ x : R_z(x, y_1, y_2, N(z_1, z_1, z_2, y_1, y_2), N(z_2, z_1, z_2, y_1, y_2)).$$

Proof. Take a recursive function $g(z, z_1, z_2, y_1, y_2, w)$ such that

$$\omega_{g(z,z_1,z_2,y_1,y_2,w)} = x : R_w(x, z, z_1, z_2, y_1, y_2, w).$$

Let h be any index of the following relation:

$$R_z(x, y_1, y_2, g(z_1, z_1, z_2, y_1, y_2, w), g(z_2, z_1, z_2, y_1, y_2, w)).$$

Then take $N(z, z_1, z_2, y_1, y_2) = g(z, z_1, z_2, y_1, y_2, h)$.

Remark. We could have proved the more general fact (call it Th. N_1) that for any r.e. relation $M(x, z, y_1, y_2, w_1, w_2)$, there is a recursive function $N(z, z_1, z_2, y_1, y_2)$ such that,

$$\omega_{N(z,z_1,z_2,y_1 y_2)} = \\ x : M(x, z, y_1, y_2, N(z_1, z_1, z_2, y_1, y_2), N(z_2, z_1, z_2, y_1, y_2)).$$

Now, in Theorem N, if we take,

$$\phi_1(z_1, z_2, y_1, y_2) = N(z_1, z_1, z_2, y_1, y_2)$$

and

$$\phi_2(z_1, z_2, y_1, y_2) = N(z_2, z_1, z_2, y_1, y_2),$$

we at once have,

Theorem 2*—A Uniform Version of Theorem 2.   *There are recursive functions* $\phi_1(z_1, z_2, y_1, y_2)$ *and* $\phi_2(z_1, z_2, y_1, y_2)$ *such that,*

$$\omega_{\phi_1(z_1, z_2, y_1, y_2)} = x : R_{z_1}(x, y_1, y_2, \phi_1(z_1, z_2, y_1, y_2), \phi_2(z_1, z_2, y_1, y_2)),$$

and

$$\omega_{\phi_2(z_1, z_2, y_1, y_2)} = x : R_{z_2}(x, y_1, y_2, \phi_1(z_1, z_2, y_1, y_2), \phi_2(z_1, z_2, y_1, y_2)).$$

Of course, Theorem 2 follows since, if a_1 and a_2 are respective indices of $M_1(x, y_1, y_2, w_1, w_2)$ and $M_2(x, y_1, y_2, w_1, w_2)$, we can take,

$$t_1(y_1, y_2) = \phi_1(a_1, a_2, y_1, y_2)$$

and

$$t_2(y_1, y_2) = \phi_2(a_1, a_2, y_1, y_2).$$

Discussion. If in the statement and proof of Theorem N, we everywhere delete "y_1" and "y_2", we get a recursive function $N(z, z_1, z_2)$ such that,

$$\omega_{N(z, z_1, z_2)} = x : R_z(x, N(z_1, z_1, z_2), N(z_2, z_1, z_2)).$$

From this, one can obtain an alternative proof of the *weak* double recursion theorem (in fact of Theorem 2.1, which is its uniform version).

For readers familiar with combinatory logic, this argument is similar to the author's proof of the double fixed point theorem in combinatory logic[2] i.e., we take a combinator N such that,

$$N z z_1 z_2 = z(N z_1 z_1 z_2)(N z_2 z_1 z_2).$$

§9.   The n-fold Recursion Theorem.   Let n be any positive integer ≥ 2. To reduce clutter, let us abbreviate "$z_1, \ldots, z_n$" by $\vec{z}$ and "$y_1, \ldots, y_n$" by $\vec{y}$. Theorem N generalizes easily as follows:

Theorem N′. *For each* $n \geq 2$ *there is a recursive function* $N(z, z_1, \ldots, z_n, y_1, \ldots, y_n)$ *such that*

$$\omega_{N(z, \vec{z}, \vec{y})} = x : R_z(x, N(z_1, \vec{z}, \vec{y}), N(z_2, \vec{z}, \vec{y}), \ldots, N(z_n, \vec{z}, \vec{y})).$$

From Theorem $N′$, we easily get the following n-fold generalization of Theorem 2.

---

[2] Smullyan [1985], Ex. 6, p. 196 and solution, p. 198

Theorem 2′. *For any r.e. relations* $M_1, \ldots, M_n$ *of* $2n + 1$ *arguments, there are recursive functions* $t_1(y_1, \ldots, y_n), \ldots, t_n(y_1, \ldots, y_n)$ *such that for each* $i \leq n$,

$$\omega_{t_i(y_1, \ldots, y_n)} = x : M_i(x, y_1, \ldots, y_n, t_1(y_1, \ldots, y_n), \ldots, t_n(y_1, \ldots, y_n)).$$

We leave the proofs of Theorem N' and Theorem 2′ to the reader. [Theorem N' has its analogue in combinatory logic. For each $n \geq 2$, there is a combinator n such that,

$$N z z_1 \ldots z_n = z(N z_1 z_1 \ldots z_n)(N z_2 z_1 \ldots z_n) \ldots (N z_n z_1 \ldots z_n).$$

This provides a simple proof of the n-fold fixed point theorem which moreover works for the λ-I calculus.]

We remark that Theorem N (and even Theorem N') is also obtainable as a corollary of the extended recursion theorem. Indeed, all the theorems of these last two chapters can be derived from the extended recursion theorem without any further diagonalizations or further use of the iteration theorem.

Exercise 7. How is Theorem N derivable from the extended recursion theorem? [Note: The corollary of Theorem E is not strong enough; we must use Theorem E for $n = 2$.]

Exercise 8. Prove Theorem N' and Theorem 2′.

§10.   A General Fixed Point Principle.

The following theorem, despite its simplicity, yields just about all the results of this chapter and the last.

Theorem G. *For each natural number* n *there is a recursive function* $g(y_1, \ldots, y_n, z)$ *with the following property: For any positive* k *and any r.e. relation* $M(x, x_1, \ldots, x_k)$ *and any recursive functions* $f_1(y_1, \ldots, y_n, z), \ldots, f_k(y_1, \ldots, y_n, z)$ *there is a number* h *such that for all* $y_1, \ldots, y_n$,

$$\omega_{g(y_1, \ldots, y_n, h)} = x : M(x, f_1(y_1, \ldots, y_n, h), \ldots, f_k(y_1, \ldots, y_n, h)).$$

Proof. (Sketch). Take a recursive g such that

$$\omega_{g(y_1, \ldots, y_n, z)} = x : R_z(x, y_1, \ldots, y_n, z).$$

Then take h to be an index of

$$M(x, f_1(y_1, \ldots, y_n, z), \ldots, f_k(y_1, \ldots, y_n, z)).$$

Let us consider some applications of Theorem G.

(1) To get the weak recursion theorem, take $n = 0$, $k = 1$,
$f_1(z) = g(z)$ and $n = g(h)$. Then $\omega_n = x : M(x, n)$.

(2) For the strong recursion theorem, take $n = 1, k = 2$,
$f_1(y, z) = y, f_2(y, z) = g(y, z)$ and $\phi(y) = g(y, h)$.

(3) For the symmetric recursion theorem, take

$$
\begin{aligned}
n &= 4, \\
k &= 5, \\
f_1(y_1, y_2, y_3, y_4, z) &= y_1; \\
f_2(y_1, y_2, y_3, y_4, z) &= y_3; \\
f_3(y_1, y_2, y_3, y_4, z) &= y_4; \\
f_4(y_1, y_2, y_3, y_4, z) &= g(y_1, y_2, y_3, y_4, z); \\
f_5(y_1, y_2, y_3, y_4, z) &= g(y_2, y_1, y_3, y_4, z); \\
t(y_1, y_2, y_3, y_4) &= g(y_1, y_2, y_3, y_4, h).
\end{aligned}
$$

Exercise 9.

1. How is Theorem N derived from Theorem G?
2. Same with the extended recursion theorem.

Chapter X

Productivity and Double Productivity

Now we will give the promised applications of the (strong) recursion and double recursion theorems to the theory of productivity and effective inseparability (and also to double productivity—a double analogue of productivity—which we will define).

I. *Productivity and Double Productivity*

§1. Weak Productivity. We recall that a set α is said to be *co-productive* under a recursive function $g(x)$ if for every number i, such that ω_i is disjoint from α, the number $g(i)$ is outside both α and ω_i. This, of course, implies the following weaker condition:

C_1: For every i, such that ω_i is disjoint from α *and such that ω_i contains at most one element*, the number $g(i)$ is outside both α and ω_i.

Condition C_1 implies the following still weaker condition:

C_2: For every i, (1) if $\omega_i = \emptyset$, then $g(i) \notin \alpha$; (2) if $\omega_i = \{g(i)\}$, then $g(i) \in \alpha$.

To see that C_1 implies C_2, suppose C_1 holds. Then (1) of C_2 is immediate. As for (2), suppose that $\omega_i = \{g(i)\}$. If $g(i) \notin \alpha$, then ω_i is a unit set disjoint from α. Hence by C_1, $g(i) \notin \omega_i$, which means that $g(i) \notin \{g(i)\}$, which is absurd. Hence $g(i) \in \alpha$.

We will say that α is *weakly* co-productive under $g(x)$ if $g(x)$ is recursive and condition C_2 holds. We wish to prove:

Theorem 1. *If α is weakly co-productive, then α is generative.*

Lemma 1. *For any recursive function $g(y)$, there is a recursive function $t(y)$ such that for all y,*

$$\omega_{t(y)} = \omega_y \cap \{gty\}.$$

Proof. We use the (strong) recursion theorem. Given the recursive function $g(y)$, let $M(x, y, z)$ be the r.e. relation $x \in \omega_y \wedge x = g(z)$. By the recursion theorem, there is a recursive function $t(y)$ such that for all y,

$$\begin{aligned} \omega_{t(y)} &= x : M(x, y, t(y)) \\ &= x : (x \in \omega_y \wedge x = gty) = (\omega_y \cap \{gty\}). \end{aligned}$$

Proof of Theorem 1. Suppose α is weakly co-productive under $g(y)$. Take $t(y)$ satisfying the above lemma. We show that gty is a generative function for α.

1. Suppose $gty \in \omega_y$. Then $\omega_y \cap \{gty\} = \{gty\}$. Also

$$\omega_{t(y)} = \omega_y \cap \{gty\}$$

(by Lemma.) Hence $\omega_{t(y)} = \{gty\}$. Hence $gty \in \alpha$ (since α is weakly co-productive under $g(y)$).
2. Suppose $gty \notin \omega_y$. Then $\omega_y \cap \{gty\} = \emptyset$. Hence $\omega_{t(y)} = \emptyset$ (since $\omega_{t(y)} = \omega_y \cap \{gty\}$). Hence $gty \notin \alpha$ (again by weak co-productivity).

By 1 and 2, $gty \in \omega_y \leftrightarrow gty \in \alpha$.

Corollary 1. *If α is weakly co-productive, then α is universal.*

Proof. By Theorem 1 above and Theorem 1 of Chapter 6.

Corollary 2—Myhill's Theorem. *If α is co-productive, then α is universal.*

Weak Co-productivity and Universality. One can go directly from weak co-productivity to universality as follows.

Lemma 1.1. *For any recursive function $g(y)$ and any r.e. set A, there is a recursive function $t(y)$ such that for all i,*

(1) $i \in A \Rightarrow \omega_{t(i)} = \{gti\}$.
(2) $i \notin A \Rightarrow \omega_{t(i)} = \emptyset$.

Proof. Apply the recursion theorem to the r.e. relation

$$y \in A \wedge x = g(z)$$

(this relation is r.e., since A is assumed r.e. and g is a recursive function). Then there is a recursive function $t(y)$ such that for all i,

$$\omega_{t(i)} = x : (i \in A \land x = gti).$$

If $i \in A$, then $\omega_{t(i)}$ is $x : x = gt(i)$, which is $\{gti\}$. If $i \notin A$, then $i \in A \land x = gti$ is false (regardless of x). Hence $\omega_{t(i)} = \emptyset$.

Now suppose that α is weakly co-productive under $g(y)$ and A is an r.e. set that we wish to reduce to α. We take $t(y)$ as in Lemma 1.1. We show that gty reduces A to α.

1. Suppose $i \in A$. Then $\omega_{t(i)} = \{gti\}$. Hence $gti \in \alpha$.
2. Suppose $i \notin A$. Then $\omega_{t(i)} = \emptyset$. Hence $gti \notin \alpha$.

§2. Weak Double Productivity.

Let us call a disjoint pair (α, β) *doubly co-productive* under a recursive function $g(x, y)$ if for all numbers i and j such that ω_i and ω_j are disjoint from each other and disjoint from α and β respectively, the number $g(i, j)$ is outside all four sets α, β, ω_i and ω_j. It is obvious that if (α, β) is doubly generative under $g(x, y)$, then (α, β) is doubly co-productive under $g(x, y)$. In T.F.S., we proved the "double" analogue of Myhill's theorem—namely that any doubly co-productive pair is D.U. We actually proved a stronger result and we are about to prove a still stronger result.

If (α, β) is doubly co-productive under $g(x, y)$, then the following weaker condition obviously holds,

D_1: For any i and j such that ω_i and ω_j are disjoint from each other and from α and β respectively *and such that $\omega_i \cup \omega_j$ contains at most one element*, the number $g(i, j)$ is outside all four sets α, β, ω_i and ω_j.

Condition D_1 implies the following condition:

D_2: For every i and j,

(1) If $\omega_i = \omega_j = \emptyset$, then $g(i, j) \notin \alpha \cup \beta$.
(2) If $\omega_i = \emptyset$ and $\omega_j = \{g(i, j)\}$, then $g(i, j) \in \beta$.
(3) If $\omega_i = \{g(i, j)\}$ and $\omega_j = \emptyset$ then $g(i, j) \in \alpha$.

We can see that D_1 implies D_2 as follows: Suppose D_1 holds. Then (1) of D_2 is obvious. As for (2), suppose $\omega_i = \emptyset$ and $\omega_j = \{g(i, j)\}$.

Then ω_i is, of course, disjoint from ω_j and α. Suppose $g(i,j) \notin \beta$. Then $\{g(i,j)\}$ is disjoint from β, which means that ω_j is disjoint from β. Hence by D_1, $g(i,j) \notin \omega_i \cup \omega_j \cup \alpha \cup \beta$, hence $g(i,j) \notin \omega_j$, and $g(i,j) \notin \{g(i,j)\}$, which is impossible. Therefore, if $\omega_i = \emptyset$ and $\omega_j = \{g(i,j)\}$, then $g(i,j) \in \beta$, which proves (2). The proof of (3) is symmetric.

We call (α, β) *weakly doubly co-productive* if α is disjoint from β and if condition D_2 is satisfied.

Theorem 2. *If* (α, β) *is weakly doubly co-productive, then* (α, β) *is doubly generative.*

We first need

Lemma 2. *For any recursive function* $g(x,y)$, *there are recursive functions* $t_1(y_1, y_2)$ *and* $t_2(y_1, y_2)$ *such that for all* i *and* j,

(1) $\omega_{t_1(i,j)} = \omega_i \cap \{g(t_1(i,j), t_2(i,j))\}$,
(2) $\omega_{t_2(i,j)} = \omega_j \cap \{g(t_1(i,j), t_2(i,j))\}$.

Proof of Lemma: Now we use the (strong) double recursion theorem—or rather its corollary Theorem 2.1, Chapter 9.

Given a recursive function $g(x,y)$, we let $M_1(x, y, z_1, z_2)$ and $M_2(x, y, z_1, z_2)$ both be the r.e. relation $x \in \omega_y \wedge x = g(z_1, z_2)$. Then by Theorem 2.1, Chapter 9, there are recursive functions $t_1(y_1, y_2)$ and $t_2(y_1, y_2)$ such that for all i and j,

1. $x \in \omega_{t_1(i,j)} \leftrightarrow (x \in \omega_i \wedge x = g(t_1(i,j), t_2(i,j)))$,
2. $x \in \omega_{t_2(i,j)} \leftrightarrow (x \in \omega_j \wedge x = g(t_1(i,j), t_2(i,j)))$.

Thus

$$\omega_{t_1(i,j)} = \omega_i \cap \{g(t_1(i,j), t_2(i,j))\}$$

(by 1) and

$$\omega_{t_2(i,j)} = \omega_j \cap \{g(t_1(i,j), t_2(i,j))\}$$

(by 2).

Proof of Theorem 2: Suppose (α, β) is weakly doubly co-productive under $g(x,y)$. Let $t_1(x,y)$ and $t_2(x,y)$ be as in the above lemma. We show that (α, β) is D.G. under the function $g(t_1(x,y), t_2(x,y))$. To show this, suppose i and j are such that ω_i and ω_j are disjoint.

1. Suppose $g(t_1(i,j), t_2(i,j)) \in \omega_i$. Then

$$\omega_i \cap \{g(t_1(i,j), t_2(i,j))\} = \{g(t_1(i,j), t_2(i,j))\}.$$

Then by the lemma,

$$\omega_{t_1(i,j)} = \{g(t_1(i,j), t_2(i,j))\}.$$

Also $g(t_1(i,j), t_2(i,j)) \notin \omega_j$ (since ω_j is disjoint from ω_i) and so

$$\omega_j \cap \{g(t_1(i,j), t_2(i,j))\} = \emptyset.$$

Hence by the lemma, $\omega_{t_2(i,j)} = \emptyset$. It then follows from (3) of condition D_2 that

$$g(t_1(i,j), t_2(i,j)) \in \alpha.$$

2. By a symmetric argument, if $g(t_1(i,j), t_2(i,j)) \in \omega_j$, then

$$g(t_1(i,j), t_2(i,j)) \in \beta.$$

3. Suppose $g(t_1(i,j), t_2(i,j)) \notin \omega_i \cup \omega_j$. Then

$$\omega_i \cap \{g(t_1(i,j), t_2(i,j))\} = \omega_j \cap \{g(t_1(i,j), t_2(i,j))\} = \emptyset,$$

and so by the lemma, $\omega_{t_1(i,j)} = \omega_{t_2(i,j)} = \emptyset$. Hence, we conclude that $g(t_1(i,j), t_2(i,j)) \notin \alpha \cup \beta$, by (1) of condition D_2.

Since ω_i is disjoint from ω_j and α is disjoint from β, it follows from 1.–3. by using propositional logic that

$$g(t_1(i,j), t_2(i,j)) \in \omega_i \leftrightarrow g(t_1(i,j), t_2(i,j)) \in \alpha$$

and

$$g(t_1(i,j), t_2(i,j)) \in \omega_j \leftrightarrow g(t_1(i,j), t_2(i,j)) \in \beta.$$

Corollary. *If (α, β) is doubly co-productive—or even weakly doubly co-productive—then (α, β) is D.U.*

Proof. By Theorem 2 above and Theorem 2 of Chapter 6.

Weak Double Co-Productivity and Double Universality. It is possible to prove the above corollary without appeal to double generativity—i.e. to pass directly from weak double co-productivity to double universality as follows.

Lemma 2.1. *For any recursive function $g(x, y)$ and any r.e. sets A and B there are recursive functions $\phi_1(y)$ and $\phi_2(y)$ such that for any number i,*

(1) $i \in A \Rightarrow \omega_{\phi_1(i)} = \{g(\phi_1(i), \phi_2(i))\}$,
(2) $i \notin A \Rightarrow \omega_{\phi_1(i)} = \emptyset$,

(3) $i \in B \Rightarrow \omega_{\phi_2(i)} = \{g(\phi_1(i), \phi_2(i))\}$,
(4) $i \notin B \Rightarrow \omega_{\phi_2(i)} = \emptyset$.

Proof. Let A and B be r.e. sets. Let $M_1(x, y, z_1, z_2)$ be the r.e. relation, $y \in A \wedge x = g(z_1, z_2)$ and let $M_2(x, y, z_1, z_2)$ be the r.e. relation, $y \in B \wedge x = g(z_1, z_2)$. Applying Theorem 2.3, Chapter 9, to M_1 and M_2 we obtain recursive functions $\phi_1(y)$ and $\phi_2(y)$, such that for all i,

(a) $\omega_{\phi_1(i)} = x : (i \in A \wedge x = g(\phi_1(i), \phi_2(i)))$,
(b) $\omega_{\phi_2(i)} = x : (i \in B \wedge x = g(\phi_1(i), \phi_2(i)))$.

(1) and (2) follow from (a); (3) and (4) follow from (b).

Now suppose (α, β) is weakly doubly co-productive under $g(x, y)$. Let (A, B) be a disjoint pair of r.e. sets that we wish to reduce to (α, β). Take recursive functions $\phi_1(y)$ and $\phi_2(y)$ as in the above lemma. We show that $g(\phi_1(y), \phi_2(y))$ reduces (A, B) to (α, β).

1. Suppose $y \in A$. Then $\omega_{\phi_1(y)} = \{g(\phi_1(y), \phi_2(y))\}$. Also we have $y \notin B$; hence $\omega_{\phi_2(y)} = \emptyset$. Therefore, $g(\phi_1(y), \phi_2(y)) \in \alpha$ (by weak double co-productivity).
2. By a symmetric argument, if $y \in B$, then $g(\phi_1(y), \phi_2(y)) \in \beta$.
3. If $y \notin A \cup B$, then $\omega_{\phi_1(y)} = \emptyset$ and $\omega_{\phi_2(y)} = \emptyset$. Therefore, we have $g(\phi_1(y), \phi_2(y)) \notin \alpha \cup \beta$ (again by weak double co-productivity).

§3.   Effective Inseparability.

We recall that a disjoint pair (α, β) is said to be E.I. under a recursive function $g(x, y)$ if for every i and j such that $\alpha \subseteq \omega_i$ and $\beta \subseteq \omega_j$ and ω_i is disjoint from ω_j, the number $g(i, j) \notin \omega_i \cup \omega_j$. We shall say that (α, β) is *weakly* E.I. under a recursive function $g(x, y)$ if α is disjoint from β and for every i and j the following conditions hold.

(1) If $\omega_i = \alpha$ and $\omega_j = \beta$, then $g(i, j) \notin \alpha \cup \beta$,
(2) If $\omega_i = \alpha$ and $\omega_j = \beta \cup \{g(i, j)\}$, then $g(i, j) \in \alpha$,
(3) If $\omega_i = \alpha \cup \{g(i, j)\}$ and $\omega_j = \beta$, then $g(i, j) \in \beta$.

Suppose (α, β) is E.I. under $g(x, y)$. Then it must be weakly E.I. under $g(x, y)$ by the following argument: Condition (1) above is immediate. As for (2), suppose $\omega_i = \alpha$ and $\omega_j = \beta \cup \{g(i, j)\}$. If $g(i, j) \notin \alpha$, then $\beta \cup \{g(i, j)\}$ is disjoint from α; hence ω_j is disjoint from ω_i; hence $g(i, j) \notin \omega_i \cup \omega_j$, which means that $g(i, j) \notin \{g(i, j)\}$, which is a contradiction. This proves (2). Condition (3) is proved by a symmetric argument.

Theorem 3. *If (α, β) is E.I.—or even weakly E.I.—and α and β are r.e., then (α, β) is D.U.*

Proof. Suppose α and β are r.e. and (α, β) is weakly E.I. under $k(x, y)$. Let $g(x, y) = k(y, x)$. Then it is obvious that (β, α) is weakly E.I. under $g(x, y)$. That is, for all x and y, (1) if $\omega_x = \beta$ and $\omega_y = \alpha$, then $g(x, y) \notin \beta$; (2) if $\omega_x = \beta$ and $\omega_y = \alpha \cup \{g(x, y)\}$, then $g(x, y) \in \beta$; (3) if $\omega_x = \beta \cup \{g(x, y)\}$ and $\omega_y = \alpha$, then $g(x, y) \in \alpha$. By the iteration theorem we can take recursive functions $t_1(y)$ and $t_2(y)$ such that for all i,

$$\omega_{t_1(i)} = \omega_i \cup \beta \quad \text{and} \quad \omega_{t_2(i)} = \omega_i \cup \alpha.$$

We show that (α, β) is weakly doubly co-productive under the function $h(x, y) = g(t_1(x), t_2(y))$.

1. Suppose $\omega_i = \omega_j = \emptyset$. Then $\omega_{t_1(i)} = \beta$ and $\omega_{t_2(j)} = \alpha$, so $g(t_1(i), t_2(j)) \notin \beta \cup \alpha$. Hence $h(i, j) \notin \alpha \cup \beta$.
2. Suppose $\omega_i = \emptyset$ and $\omega_j = \{h(i, j)\} = \{g(t_1(i), t_2(j))\}$. Then $\omega_{t_1(i)} = \beta$ and $\omega_{t_2(j)} = \alpha \cup \{g(t_1(i), t_2(j))\}$. Hence

$$h(i, j) = g(t_1(i), t_2(j)) \in \beta$$

   by (2), with $x = t_1(i)$ and $y = t_2(j)$.
3. By a symmetric argument, if $\omega_i = \{g(t_1(i), t_2(i))\}$ and $\omega_j = \emptyset$, then $g(t_1(i), t_2(j)) \in \alpha$.

Since (α, β) is weakly doubly co-productive, (α, β) is doubly universal by Theorem 2, Corollary 1.

Exercise 1. It is possible to prove Theorem 3 without appeal to Theorem 2 or its corollary—that is, we can pass directly from weak E.I. + r.e. to D.U. as follows.

Step 1. We need the result of Exercise 5 of Chapter 9. [To establish this result, let $M_1(x, y, z)$ be the r.e. relation $x \in \alpha \vee (y \in B \wedge x = z)$ and let $M_2(x, y, z)$ be the r.e. relation $x \in \beta \vee (y \in A \wedge x = z)$. Then apply Theorem 2.4, Chapter 9 to the relations M_1 and M_2.]

Step 2. Suppose (α, β) is weakly E.I. under $g(x, y)$. Let (A, B) be a disjoint pair of r.e. sets that we wish to reduce to (α, β). Take $\phi_1(y)$ and $\phi_2(y)$ as in Step 1 and show that the function $g(\phi_1(y), \phi_2(y))$ reduces (A, B) to (α, β).

§4. **A Retrospective Look.** Historically, the first proofs of the Ehrenfeucht-Feferman and the Putnam-Smullyan theorems were based on creativity and effective inseparability rather than complete creativity and complete effective inseparability. Now that we have proved Theorems 1 and 3 of this chapter, we can give the original arguments.

The Ehrenfeucht-Feferman argument (using our terminology) was this: Suppose S is a consistent axiomatizable extension of (R). Then, (1) S is a Rosser system for sets; (2) all recursive functions of one argument are strongly definable in S. By (1), some E.I. pair (A, B) is strongly separated in S by some formula $H(v_1)$. Then $H(v_1)$ represents some r.e. superset A' of A disjoint from B. Then the pair (A', B) is obviously also E.I., hence A' is a creative set (Exercise 8, Chapter 5). Then by Myhill's theorem, A' is universal. Thus, some universal set is representable in S. Then by (2), all r.e. sets are representable in S.

In the Putnam-Smullyan argument, again some E.I. pair (A, B) is strongly separated in S by some formula $H(v_1)$; this formula *exactly* separates a disjoint pair (A', B') of r.e. supersets of A and B. Then (A', B') is obviously E.I. and since A' and B' are r.e., (A', B') is D.U. by Theorem 3. Thus some D.U. pair is exactly separable in S. Hence by (2), every disjoint pair of r.e. sets is exactly separable in S.

As we have said, these were historically the first arguments. They used the recursion and double recursion theorems. Then came Shepherdson's arguments which used what we might call "Shepherdsonian self-reference" in place of recursion theorems. And now, as we have demonstrated in this volume, by using the notions of complete creativity and complete effective inseparability, we can achieve the same metamathematical results (about consistent axiomatizable extensions of (R)) without using either recursion theorems or Shepherdsonian self-reference. All three approaches strike us as equally interesting.

Of course, the results of this chapter also provide alternative proofs of Theorems A and B of Chapter 6.

Exercise 2. Explain why the last statement is true.

Chapter XI

Three Special Topics

The topics of this chapter are of more specialized interest and are not necessary for the results of our final chapter. They will probably be of more interest to the specialist (particularly the results of Section III) than to the general reader.

I. Uniform Reducibility

We know from Chapter 2 that if all *recursive* sets are representable in S, then S is undecidable. We also know from Chapter 4 that if all recursively enumerable sets are representable in S, then S is not only undecidable but generative. We might also ask the question, If all recursive sets are representable in S, is S necessarily generative? Shoenfield [1961] answered this question negatively. He constructed an axiomatizable system in which all recursive sets are representable, and so the system is undecidable, but he showed that the system is not creative.

Let us say that all recursive sets are *uniformly* representable in S if there is a recursive function $g(x)$ such that for any number i, if ω_i is a *recursive* set, then $g(i)$ is the Gödel number of a formula which represents ω_i in S. We will show that if all recursive sets are *uniformly* representable in S, then S is generative.[1]

---

[1] We proved this in T.F.S. using Theorem 1, Chapter 10, whose proof in turn uses the recursion theorem. We gave an alternative proof in Smullyan [1963] based on Theorem 1*, Chapter 6, which does not use the recursion theorem. The proof we give in this chapter is simpler still. We sketch the original proof in some of the exercises.

We also showed in Chapter 2 that if all recursive sets are *definable* in S and S is consistent, then the pair (P, R) of its nuclei is recursively inseparable. Under the same hypothesis, is the pair (P, R) necessarily effectively inseparable? The answer is *no*. In Shoenfield's system all recursive sets are not only representable, but definable. However, the set P is not creative. Hence the pair (P, R) of nuclei of the system, though recursively inseparable, is not effectively inseparable.

Let us say that all recursive sets are *uniformly* definable in S if there is a recursive function $g(x, y)$ such that for any numbers i and j, if ω_j is the complement of ω_i, then $g(i, j)$ is the Gödel number of a formula which defines ω_i in S. We will show that if all recursive sets are uniformly definable in S and S is consistent, then the pair (P, R) is completely effectively inseparable (in fact semi-D.U.). The proof will be based on (a) of Theorem 2*, Chapter 6.

Remark. Informally speaking, to say that all recursive sets are uniformly definable in S is to say that given any index of a recursive set α *as well as an index of its complement*, we can find a formula which defines α in S. The reader might wonder why we require that we be given an index of the complement of α as well as an index of α. The fact is that there is no consistent system S with the stronger property that there exists a recursive function $f(x)$ such that for every number i for which ω_i is recursive, $f(i)$ is the Gödel number of a formula which defines ω_i in S. We are indebted to John Myhill for this observation.

§1. Uniform Reducibility.

We will say that a collection C of r.e. sets is *uniformly reducible* to a set α if there is a recursive function $f(x, y)$—which we call a *uniform reduction* of C to α—such that for every i for which $\omega_i \in C$, the function $f(i, x)$ (as a function of x) reduces ω_i to α.

Lemma 1. *If $\{N, \emptyset\}$ is uniformly reducible to α, then α is generative relative to $\{N, \emptyset\}$.*

Proof. Suppose $f(x, y)$ is a uniform reduction of $\{N, \emptyset\}$ to α. We show that the function $f(x, x)$ is a generative function for α relative to $\{N, \emptyset\}$.

1. Suppose $\omega_i = N$. Then for all x, $x \in N \leftrightarrow f(i, x) \in \alpha$. Hence for all x, $f(i, x) \in \alpha$, so $f(i, i) \in \alpha$.

2. Suppose $\omega_i = \emptyset$. Then for all x, $x \in \emptyset \leftrightarrow f(i,x) \in \alpha$. Hence for all x, $f(i,x) \notin \alpha$, so $f(i,i) \notin \alpha$.

By 1 and 2, and Proposition 1, Chapter 6, α is generative relative to $\{N, \emptyset\}$ under the function $f(x,x)$.

Proposition 1. *If the collection $\{N, \emptyset\}$ is uniformly reducible to α, then α is universal.*

Proof. By the above lemma and Theorem 1*, Chapter 6.

Theorem 1. *If the collection of all recursive sets is uniformly reducible to α, then α is universal.*

Proof. Immediate from Proposition 1.

Corollary. *If all recursive sets are uniformly representable in S, then S is generative.*

Proof. Suppose $g(x)$ is a recursive function such that for all i for which ω_i is recursive, $g(i)$ is the Gödel number of a formula that represents ω_i in S. Then the function $r(g(x),y)$ uniformly reduces the collection of recursive sets to P. Then by Theorem 1, P is universal and, hence, generative.

Remark. We see from the above arguments that, in fact, a sufficient condition for S to be generative is that the collection $\{N, \emptyset\}$ be uniformly representable in S.

Exercise 1. Prove the following generalization of Lemma 1: Suppose that a collection C is uniformly reducible to α under $f(x,y)$ and that for every i such that $\omega_i \in C$, the set of all x such that $f(x,x) \in \omega_i$ is also in C. Then α is generative relative to C—more specifically, if $t(y)$ is an iterative function for the relation $f(x,x) \in \omega_y$, so that $x \in \omega_{t(y)}$ iff $f(x,x) \in \omega_y$, then α is generative relative to C under the function $f(t(x),t(x))$.

Exercise 2. How does Lemma 1 follow from the result of Exercise 1?

Exercise 3. Using Exercise 1, show that if the collection of all recursive sets is uniformly reducible to α, then α is generative relative to that collection.

Exercise 4. Show the trivial fact that if α is generative relative to the collection of all recursive sets, then α is weakly co-productive.

Exercise 5. Using the last two exercises, show that Corollary 1 of Theorem 1, Chapter 10, yields an alternative proof of Theorem 1 above.

Exercise 6. Let C_1 be the collection consisting of the empty set and all sets with exactly one element. Show that if C_1 is uniformly reducible to α under a 1-1 function $f(x, y)$, then α is universal. [Hint: Show that C_1 and $f(x, y)$ then satisfy the hypothesis of Exercise 1.]

§2. Uniform Reducibility for Pairs.

We shall say that a collection $\mathcal{D}$ of disjoint ordered pairs of r.e. sets is uniformly reducible to a disjoint pair (α, β) under a recursive function $g(x, y, z)$ if for every i and j such that $(\omega_i, \omega_j) \in \mathcal{D}$, the function $g(i, j, x)$ reduces (ω_i, ω_j) to (α, β).

Lemma 2. If the collection $\mathcal{D}_2$ (viz. $\{(N, \emptyset), (\emptyset, N)\}$) is uniformly reducible to (α, β), then (α, β) is D.G. relative to $\mathcal{D}_2$.

Proof. Suppose $\mathcal{D}_2$ is uniformly reducible to (α, β) under $g(x, y, z)$. We show that $g(x, y, y)$ is a D.G. function for (α, β) relative to $\mathcal{D}_2$.

Suppose $\omega_i = N$ and $\omega_j = \emptyset$. Then $g(i, j, x)$ reduces $(N, \emptyset)$ to (α, β), so for all x, $x \in N \leftrightarrow g(i, j, x) \in \alpha$ and $x \in \emptyset \leftrightarrow g(i, j, x) \in \beta$, which means that $g(i, j, x) \in \alpha$ and $g(i, j, x) \notin \beta$, so $g(i, j, x) \in \alpha - \beta$. Therefore, $g(i, j, j) \in \alpha - \beta$.

Similarly, if $\omega_i = \emptyset$ and $\omega_j = N$, then $g(i, j, j) \in \beta - \alpha$. Then by statement (1), Proposition 2, Chapter 6, (α, β) is D.G. relative to $\mathcal{D}_2$ under $g(x, y, y)$.

Remark. The function $g(x, y, x)$ would have served as well—in fact for any constant c, the function $g(x, y, c)$ would work.

From Lemma 2 and (a) of Theorem 2*, Chapter 6, we have

Proposition 2. If $\mathcal{D}_2$ is uniformly reducible to (α, β), then (α, β) is semi-D.U.

Theorem 2. If the collection of all complementary pairs $(A, \widetilde{A})$ of r.e. sets is uniformly reducible to (α, β), then (α, β) is semi-D.U.

Corollary. If all recursive sets are uniformly definable in $\mathcal{S}$ and $\mathcal{S}$ is consistent, then the pair (P, R) of nuclei of $\mathcal{S}$ is semi-D.U. (and, hence, completely E.I.).

Proof of Corollary. Suppose $\mathcal{S}$ is consistent and that $g(x, y)$ is a recursive function such that for any recursive set ω_i, if ω_j is the complement of ω_i, then $g(i, j)$ is the Gödel number of a formula which defines ω_i in $\mathcal{S}$. By the assumption of consistency, the formula completely represents ω_i in S and, therefore, exactly separates the pair (ω_i, ω_j) in S. From this it follows that the function $r(g(x,y), z)$ is a uniform reduction of the collection of all complementary pairs of recursive sets to the pair (P, R). Then by Theorem 2, the pair (P, R) is semi-D.U. (and, hence, completely E.I.).

The exercises that follow concern the notion of uniform definability of recursive *functions* in a system S.

Exercise 7. We shall call a number i an (r.e.) index of a recursive function $f(x)$ if i is an index of the relation $f(x) = y$.

Show that there is a recursive function $\phi(x)$ such that for any number i, if i is an index of a recursive function $f(x)$, then $\phi(i)$ is an index of the relation $f(x) \neq y$.

Exercise 8. Let us say that all recursive functions of one argument are *uniformly* definable in S if there is a recursive function $g(x)$ such that for any number i, if i is an index of a recursive function $f(x)$, then $g(i)$ is the Gödel number of a formula that defines $f(x)$ in S.

Show that if S is effectively a Rosser system for binary relations, then all recursive functions of one argument are uniformly definable in S. [We suggest you use the last exercise.]

Exercise 9. Prove that there is a recursive function $t(x, y)$ such that for any numbers i and j, if ω_i and ω_j are complementary, then $t(i, j)$ is an index of the characteristic function of ω_i.

Exercise 10. Show that there is a recursive function $\phi(x)$ such that for any number i and any set A, if i is the Gödel number of a formula that defines in S the characteristic function of A, then $\phi(i)$ is the Gödel number of a formula that defines A in S.

Exercise 11. Using the last two exercises and Theorem 2, prove that if all recursive functions of one argument are uniformly definable in S and S is consistent, then S is a completely E.I. system.

II. Pseudo-uniform Reducibility

In Ch. 2 we showed

Theorem C_1 (Th. 4, Ch. 2). *If all recursive sets are representable in a system S, then S is undecidable.*

Theorem C_2 (Th. 13, Ch. 2). *If all recursive sets are definable in S and S is consistent, then not only is S undecidable, but the pair (P, R) is recursively inseparable.*

The above theorems combine notions of recursive function theory with those of first-order systems. We shall now obtain generalizations of them of a purely recursive function theoretic nature and we will also show that the conclusions of Theorems C_1 and C_2 hold under weaker hypotheses. The results that follow are those of Smullyan [1963A].

§3. Pseudo-uniform Reducibility.

We know that if every r.e. set is reducible to a set α, then α must be non-recursive (in fact even generative). Suppose that every *recursive* set is reducible to α. Does it necessarily follow that α must be non-recursive? Certainly not, for if α is any non-empty set whose complement is non-empty, then every recursive set A is reducible to α as follows: Take any number a_1 in α and any number a_2 not in α and define $g(x) = a_1$ if $x \in A$, and $g(x) = a_2$ if $x \notin A$. Since A is recursive, so is the function $g(x)$ (why?) and for all x, $x \in A \leftrightarrow g(x) \in \alpha$, so $g(x)$ reduces A to α. Thus, it is not true that if all recursive sets are reducible to α, then α must be non-recursive.

On the other hand, as we saw in Part I of this chapter, if the collection of all recursive sets is *uniformly* reducible to α, then α is not only non-recursive, but even generative. Thus, to establish the non-recursivity of a set α, the hypothesis that all recursive sets are reducible to α is too weak, whereas the hypothesis of uniform reducibility is stronger than necessary (as we will see). And so we turn to a notion of intermediate strength.

We will say that a collection C of number sets is *pseudo-uniformly reducible* to α if there is a recursive function $g(x, y)$—which we call a pseudo-uniform reduction of C to α—such that for any set A in C, there is some number i such that $g(i, x)$ (as a function of x) reduces A to α (i.e. for all x, $x \in A \leftrightarrow g(i, x) \in \alpha$). If C is a collection of r.e. sets, our definition does not require that for *any* index i of a member A of C, $g(i, x)$ reduces A to α, but only that there is *some* number i (not necessarily even an index of A) such that $g(i, x)$ reduces A to α. Thus, the notion of pseudo-uniform reducibility is much weaker than uniform reducibility. We shall now show that if the collection of all recursive sets is pseudo-uniformly reducible to α, then α must be non-recursive. We first show the following stronger fact.

Theorem 3. *A sufficient condition for a set α to be non-recursive is that there is a recursive function $f(x)$ such that for every recursive set A, there is at least one number i such that $i \in A \leftrightarrow f(i) \in \alpha$.*

Proof. Suppose that the condition holds. Consider any recursive set A. Then $f^{-1}(A)$ is recursive. Hence, there is a number i such that $i \in f^{-1}(A) \leftrightarrow f(i) \in \alpha$. Hence, $f(i) \in A \leftrightarrow f(i) \in \alpha$. This shows that A cannot be the complement $\tilde{\alpha}$ of α and, hence, $\tilde{\alpha}$ cannot be recursive, so α cannot be recursive.

As a corollary we have

Theorem 4. *If the collection of all recursive sets is pseudo-uniformly reducible to α, then α is non-recursive.*

Proof. Suppose $g(x,y)$ is a pseudo-uniform reduction of all recursive sets to α. Let $f(x)$ be the recursive function $g(x,x)$. Then for any recursive set, there is a number i such that for all x,

$$x \in A \leftrightarrow g(i,x) \in \alpha.$$

Hence $i \in A \leftrightarrow g(i,i) \in \alpha$, and so $i \in A \leftrightarrow f(i) \in \alpha$. Then by Theorem 3, α is non-recursive.

Metamathematical Applications. Theorem 4 contains the essential mathematical essence of Theorem C_1, for suppose all recursive sets are representable in S. We consider the representation function $r(x,y)$ of S ($r(i,j)$ is the Gödel number of $E_i[j]$). Then $r(x,y)$ is obviously a pseudo-uniform reduction of the collection of all recursive sets to the set P, and so by Theorem 4, the set P cannot be recursive (which means that S is undecidable).

From Theorem 3, however, we get the following apparently stronger result.

Theorem C_1^*. *Suppose that for every recursive set A there is a number i that is the Gödel number of a formula $F_i(v_1)$ such that $F_i(\bar{i})$ is provable in S iff $i \in A$. Then S is undecidable.*

Proof. Assume hypothesis. Then for every recursive set A there is a number i such that $d(i) \in P \leftrightarrow i \in A$, where $d(x)$ is the diagonal function $r(x,x)$. Then P is non-recursive by Theorem 3.

§4.   Pseudo-uniform Reducibility for Pairs.   We now consider a collection $\mathcal{D}$ of ordered pairs of r.e. sets. We shall say that $\mathcal{D}$ is *pseudo-uniformly reducible* to a pair (α, β) of number sets if

there is a recursive function $g(x, y)$—which we call a pseudo-uniform reduction of $\mathcal{D}$ to (α, β)—such that for every pair (A, B) in $\mathcal{D}$, there is a number i such that $g(i, x)$ (as a function of x) reduces (A, B) to (α, β). Let us note that if $h(x, y, z)$ is a uniform reduction of $\mathcal{D}$ to (α, β) (as defined in Ch. 6) and if we take $g(x, y)$ to be the function $h(Kx, Lx, y)$, then $g(x, y)$ is a pseudo-uniform reduction of $\mathcal{D}$ to (α, β) (because for any pair (ω_i, ω_j) in $\mathcal{D}$, $h(i, j, y)$ reduces (ω_i, ω_j) to (α, β), but $h(i, j, y) = g(J(i, j), y)$, and so $g(c, y)$ reduces (ω_i, ω_j) to (α, β), where $c = J(i, j)$). Thus, pseudo-uniform reducibility (for pairs) is a weaker notion than uniform reducibility.

Theorem 5. *A sufficient condition for a disjoint pair (α, β) to be recursively inseparable is that there is a recursive function $f(x)$ such that for every complementary pair (A, B) of recursive sets $(B = \tilde{A})$, there is a number i such that $i \in A \leftrightarrow f(i) \in \alpha$, and $i \in B \leftrightarrow f(i) \in \beta$.*

Proof. Assume hypothesis. Suppose (α, β) is recursively separable. Then there is a complementary pair (A, B) of recursive supersets of α and β respectively. Then $(f^{-1}(B), f^{-1}(A))$ is a complementary pair of recursive sets, and so there is a number i such that

$$i \in f^{-1}(B) \leftrightarrow f(i) \in \alpha,$$

and

$$i \in f^{-1}(A) \leftrightarrow f(i) \in \beta.$$

Hence $f(i) \in B \leftrightarrow f(i) \in \alpha$, and $f(i) \in A \leftrightarrow f(i) \in \beta$. This is clearly impossible, since $\alpha \subseteq A$ and $\beta \subseteq B$ and A is the complement of B. Therefore, (α, β) cannot be recursively separable.

Next we note that if $g(x, y)$ is a pseudo-uniform reduction of the collection of all complementary pairs of recursive sets to (α, β) (where α and β are disjoint), then the hypothesis of Theorem 5 holds, taking $f(x) = g(x, x)$, and we have

Theorem 6. *If the collection of all complementary pairs of recursive sets is pseudo-uniformly reducible to (α, β) (where α and β are disjoint) then (α, β) is recursively inseparable.*

Metamathematical Applications. Theorem C_2 is but a special case of Theorem 6, because if all recursive sets are definable in $\mathcal{S}$ and $\mathcal{S}$ is consistent, then all recursive sets are completely representable in $\mathcal{S}$. Hence $r(x, y)$ is a pseudo-uniform reduction of the collection of all complementary pairs of recursive sets to the pair (P, R).

From Theorem 5, however, we get the apparently stronger result.

Theorem C$_2^*$. A sufficient condition for the nuclei (P, R) of a system S to be recursively inseparable is that for every recursive set A, there is a number i that is the Gödel number of a formula $F_i(v_1)$ such that $F_i(\bar{i})$ is provable in S iff $i \in A$, and $F_i(\bar{i})$ is refutable in S iff $i \notin A$.

III.　Some Feeble Partial Functions

In this section we will obtain some curious strengthenings of some earlier results.

Let $\psi(x_1, \ldots, x_n)$ be a function defined on some, but not necessarily all, n-tuples of natural numbers. We call ψ a *partial recursive function* if the relation $\psi(x_1, \ldots, x_n) = y$ (i.e., the set of all $(n + 1)$-tuples $(x_1, \ldots, x_n, y)$ such that ψ is defined on $(x_1, \ldots, x_n)$ and assigns it the value y) is an r.e. relation. [Unlike the case of total recursive functions, this does not, in general, imply that the relation $\psi(x_1, \ldots, x_n) = y$ is recursive. It does, however, if the domain of ψ happens to be recursive.]

The results of this section are about partial recursive functions.

§5.　Feeble Co-productive and Generative Functions.
It is well known (cf., e.g., Rogers [1967]) that a sufficient condition for a set α to be co-productive is that there exists a *partial* recursive function $\psi(x)$ such that for all i, if ω_i is disjoint from α, then ψ is defined on i and $\psi(i) \notin \alpha \cup \omega_i$. We will show that the hypothesis can be simultaneously weakened in two ways, (1) Except for the case that $\omega_i = \emptyset$, it is not necessary to assume that if ω_i is disjoint from α, then ψ is defined on i, but only that if ω_i is disjoint from α *and if ψ happens to be defined on i*, then $\psi(i) \notin \omega_i$; (2) moreover, this need hold only when ω_i happens to be the set $\{\psi(i)\}$. This leads to the following definition.

Definition 1. α is *feebly* co-productive under a partial recursive function $\psi(x)$ iff for every i, the following two conditions hold:

(1) If $\omega_i = \emptyset$, then ψ is defined on i and $\psi(i) \notin \alpha$.
(2) If ψ is defined on i and $\omega_i = \{\psi(i)\}$, then $\psi(i) \in \alpha$.

We will prove

Theorem 7. *If α is feebly co-productive (under some partial recursive function), then α is universal.*

We will also state and prove a strengthening of Theorem 2*, (b), Chapter 6.

Definition 2. α is *feebly generative* under a partial recursive function $\psi(x)$ iff for every i, the following two conditions hold:

(1) If $\omega_i = \emptyset$, then ψ is defined on i and $\psi(i) \notin \alpha$.
(2) If $\omega_i = N$ and if ψ happens to be defined on i, then $\psi(i) \in \alpha$.

Theorem 8. *If α is feebly generative (under some partial recursive function), then α is universal.*

The proofs of both Theorem 7 and Theorem 8 utilize the recursion theorem and bring to light Theorem I below (which we find to be of particular interest). Let us call a (total) recursive function $h(y)$ an *associate* of a partial recursive function $\psi(y)$ if the following two conditions hold:

1. For every number i, ψ is defined on $h(i)$ — i.e., the function $\psi h y$ is total.
2. For every number i, $\omega_{h(i)} = \omega_i$.

Theorem I. *If $\psi(y)$ is defined on all indices of the empty set, then $\psi(y)$ has an associate.*

We first prove two lemmas.

Lemma 3. *For any r.e. set A, there is a recursive function $h(y)$ such that for all y,*

(1) *If $h(y) \in A$, then $\omega_{h(y)} = \omega_y$.*
(2) *If $h(y) \notin A$, then $\omega_{h(y)} = \emptyset$.*

Proof. Given an r.e. set A, let $M(x,y,z)$ be the r.e. relation

$$x \in \omega_y \wedge z \in A.$$

By the recursion theorem, there is a recursive function $h(y)$ such that for all y, $\omega_{h(y)} = x : M(x,y,h(y))$, so

$$\omega_{h(y)} = x : (x \in \omega_y \wedge h(y) \in A).$$

If $h(y) \in A$, then $\omega_{h(y)} = x : x \in \omega_y = \omega_y$; if $h(y) \notin A$, then $x \in \omega_y \wedge h(y) \in A$ is false for all x, hence $\omega_{h(y)} = \emptyset$.

Lemma 4. *If A is r.e. and contains all indices of the empty set, then there is a recursive function $h(y)$ such that for all y the following two* *conditions hold:*

(1) $\omega_{h(y)} = \omega_y$,
(2) $h(y) \in A$.

Proof. Assume hypothesis. Let $h(y)$ be as in Lemma 3. Suppose $h(y) \notin A$. Then by (2) of Lemma 3, $\omega_{h(y)} = \emptyset$. Hence $h(y)$ is an index of $\emptyset$ and, hence, $h(y) \in A$ by hypothesis. This is a contradiction. So for all y, $h(y) \in A$, which proves (2). Then by (1) of Lemma 3, $\omega_{h(y)} = \omega_y$ for all y, which proves (1) above.

Proof of Theorem I. Suppose $\psi(y)$ is a partial recursive function defined on at least all indices of the empty set. Let A be the domain of ψ (the set of numbers on which $\psi(y)$ is defined). Then A is r.e. and A contains all indices of the empty set. Take $h(y)$ satisfying Lemma 3. Then for all y, $\omega_{h(y)} = \omega_y$. Also for all y, $h(y) \in A$, which means ψ is defined on $h(y)$. Therefore, $h(y)$ is an associate of $\psi(y)$.

Proof of Theorem 7. Suppose α is feebly co-productive under $\psi(y)$. By Theorem I, $\psi(y)$ has an associate $h(y)$. For any number i,

1. Suppose $\omega_i = \emptyset$. Then $\omega_{h(i)} = \emptyset$. Since ψ is defined on $h(i)$, then $\psi hi \notin \alpha$ (since α is feebly co-productive under ψ).
2. Suppose $\omega_i = \{\psi hi\}$. Then $\omega_{h(i)} = \{\psi hi\}$. Hence $\psi hi \in \alpha$ (since α is feebly co-productive under ψ).

By 1 and 2, α is weakly co-productive under ψhy. Then by Corollary 1 of Theorem 1, Chapter 10, α is universal.

Proof of Theorem 8. Suppose α is feebly generative under $\psi(y)$. Since ψ is defined on all indices of the empty set, it has an associate $h(y)$ (by Theorem I). For any i,

1. Suppose $\omega_i = \emptyset$. Then $\omega_{h(i)} = \emptyset$. Hence $\psi hi \notin \alpha$.
2. Suppose $\omega_i = N$. Then $\omega_{h(i)} = N$. Also ψ *is* defined on $h(i)$.

Hence $\psi hi \in \alpha$.

By 1 and 2, α is generative relative to $\{\emptyset, N\}$ under ψhy. The conclusion follows by Theorem 1*, Chapter 6.

§6. Double Analogues.

All the results of §5 have double analogues. In what follows, $\psi(x, y)$ is a partial recursive function of two variables.

Definition 3. A disjoint pair (α, β) is *feebly* doubly co-productive under $\psi(x, y)$ iff for all i and j, the following conditions hold:

(1) If $\omega_i = \omega_j = \emptyset$, then ψ is defined on (i, j) and $\psi(i, j) \notin \alpha \cup \beta$.
(2) If ψ is defined on (i, j) and $\omega_i = \emptyset$ and $\omega_j = \{\psi(i, j)\}$, then $\psi(i, j) \in \beta$.
(3) If ψ is defined on (i, j) and $\omega_i = \{\psi(i, j)\}$ and $\omega_j = \emptyset$, then $\psi(i, j) \in \alpha$.

Definition 4. A disjoint pair (α, β) is *feebly* D.G. under $\psi(x, y)$ iff for all i and j, the following conditions hold:

(1) If $\omega_i = \omega_j = \emptyset$, then ψ is defined on (i, j) and $\psi(i, j) \notin \alpha \cup \beta$.
(2) If $\omega_i = N$ and $\omega_j = \emptyset$ and if ψ is defined on (i, j), then

$$\psi(i, j) \in \alpha.$$

(3) If $\omega_i = \emptyset$ and $\omega_j = N$ and if ψ is defined on (i, j), then

$$\psi(i, j) \in \beta.$$

We will prove

Theorem 9. *If (α, β) is feebly doubly co-productive (under some partial recursive function $\psi(x, y)$), then (α, β) is D.U.*

Theorem 10. *If (α, β) is feebly D.G. (under some partial recursive function $\psi(x, y)$), then (α, β) is D.U.*

To prove Theorems 9 and 10, we need to introduce a "double" analogue of the notion of an *associate*. By a *double associate* of a partial recursive function $\psi(x, y)$, we shall mean an (ordered) pair $(h_1(x, y), h_2(x, y))$ of (total) recursive functions such that the following two conditions hold:

1. The function $\psi(h_1(x, y), h_2(x, y))$ is total.
2. For all numbers i and j, $\omega_{h_1(i,j)} = \omega_i$ and $\omega_{h_2(i,j)} = \omega_j$.

In place of Theorem I, we now need

Theorem II. *If for all indices i and j of the empty set, ψ is defined on (i, j), then ψ has a double associate.*

Again we shall first prove two lemmas.

Lemma 5. *For any r.e. relation $R(y_1, y_2)$, there are recursive functions $h_1(y_1, y_2)$ and $h_2(y_1, y_2)$ such that for all y_1 and y_2,*

(1) $\omega_{h_1(y_1, y_2)} = x : (x \in \omega_{y_1} \wedge R(h_1(y_1, y_2), h_2(y_1, y_2)))$,
(2) $\omega_{h_2(y_1, y_2)} = x : (x \in \omega_{y_2} \wedge R(h_1(y_1, y_2), h_2(y_1, y_2)))$.

Proof. Let $M_1(x, y, z_1, z_2)$ be the relation $x \in w_y \wedge R(z_1, z_2)$. Let $M_2(x, y, z_1, z_2)$ be the same relation. Then apply Theorem 2.1, Chapter 9.

Lemma 6. *For any r.e. relation $R(y_1, y_2)$, if for all i and j such that $\omega_i = \omega_j = \emptyset$ and $R(i,j)$ holds, then there are recursive functions $h_1(y_1, y_2)$ and $h_2(y_1, y_2)$ such that for all y_1 and y_2,*

(1) $R(h_1(y_1, y_2), h_2(y_1, y_2))$.
(2) $\omega_{h_1(y_1,y_2)} = \omega_{y_1}$ and $\omega_{h_2(y_1,y_2)} = \omega_{y_2}$.

Proof. Assume hypothesis. Take recursive functions $h_1(y_1, y_2)$ and $h_2(y_1, y_2)$ satisfying Lemma 5.

Suppose $R(h_1(y_1, y_2), h_2(y_1, y_2))$ doesn't hold. Then by (1) and (2) of Lemma 5, $\omega_{h_1(y_1,y_2)}$ and $\omega_{h_2(y_1,y_2)}$ will both be empty. Hence $R(h_1(y_1, y_2), h_2(y_1, y_2))$ will hold (by hypothesis), which is a contradiction. Hence, $R(h_1(y_1, y_2), h_2(y_1, y_2))$ must hold. By (1) and (2) of Lemma 5, we see that $\omega_{h_1(y_1,y_2)} = \omega_{y_1}$ and $\omega_{h_2(y_1,y_2)} = \omega_{y_2}$.

Proof of Theorem II. Assume hypothesis. Define $R(x, y)$ iff ψ is defined on (x, y). Then R is r.e. and satisfies the hypothesis of Lemma 6. The conclusion easily follows by Lemma 6.

Using Theorem II, the proofs of Theorems 9, 10 are obvious modifications of the proofs of Theorems 7 and 8 and are left as exercises.

Chapter XII

Uniform Gödelization

We conclude this volume with some pretty applications of recursion and double recursion theorems and some variants of Shepherdson's arguments. We obtain a substantial strengthening of Shepherdson's theorem as well as some new results on *uniform incompletability*, which we now define.

Given a consistent axiomatizable system S, we let S_n be that system whose axioms are those of S together with all formulas whose Gödel number is in ω_n. [We might refer to S_n as the nth extension of S.] We call S *uniformly incompletable* if there is a formula $H(v_1)$ such that for any n for which S_n is consistent, $H(\overline{n})$ is an undecidable sentence of S_n. [In a sense, for each n for which S_n is consistent, $H(\overline{n})$ can be thought of as asserting its own non-provability in S_n— or more accurately that it is not provable in S_n before it is refutable in S_n.] Marian Pour-El [1968] proved that every consistent axiomatizable extension of (R) is uniformly incompletable (a part of her argument is a variant of the proof of the Putnam-Smullyan theorem). We have been independently working on this problem along entirely different lines which reveal that such systems possess some interesting properties apparently stronger than uniform incompletability. It is to these stronger properties that we first turn.

I. The Sentential Recursion Property

We shall say that S has the sentential recursion property if for every r.e. relation $R(x,y)$ there is a number h such that $x : R(x,h)$ is represented in S by a formula $H(v_1)$ whose Gödel number is h. Using the *weak* recursion theorem, we will prove a result that implies that

141

every consistent axiomatizable extension of (R) has the sentential recursion property.

§1. Effectively Gödelian Systems.

In T.F.S. we termed a system S *Gödelian* if all r.e. sets are representable in S. We shall now say that S is *effectively* Gödelian if all r.e. sets are uniformly representable in S as defined in Ch. 4—i.e. if there is a recursive function $g(x)$ (under which S will be said to be effectively Gödelian) such that for every number i, $g(i)$ is the Gödel number of a formula which represents ω_i in S.

Theorem 1. *If S is effectively Gödelian, then S has the sentential recursion property.*

Proof. Suppose S is effectively Gödelian under $g(x)$. Now consider any r.e. relation $R(x, y)$. By Theorem 1.1, Ch. 8, there is a number k such $\omega_k = x : R(x, g(k))$. But $g(k)$ is the Gödel number of a formula that represents ω_k and, hence, represents $x : R(x, g(k))$. And so we take $h = g(k)$.

We proved in Ch. 4 (Proposition 5, Corollary) that every consistent axiomatizable extension of (R) is effectively Gödelian. And so by Theorem 1 we have

Theorem R_1. *Every consistent axiomatizable extension of (R) has the sentential recursion property.*

Some consequences of the Sentential Recursion Property. For any number n, let us define W_n as the set of all *expressions of S whose Gödel number is in ω_n.*

Theorem 1.1. *If S has the sentential recursion property, then for any recursive function $f(x)$, there is a formula $H(v_1)$ such that for every number n, $H[\overline{n}]$ is provable in S if, and only if, $H[\overline{n}] \in W_{f(n)}$.*

Proof. Suppose S has the sentential recursion property. We let $r(x, y)$ be the Gödel number of $E_x[\overline{y}]$ (as usual). Now, given a recursive function $f(x)$, we take $R(x, y)$ to be the r.e. relation $r(y, x) \in \omega_{f(x)}$. By hypothesis there is a formula $H(v_1)$ with Gödel number h which represents $x : R(x, h)$. Then $H(v_1)$ represents $x : r(h, x) \in \omega_{f(x)}$, and so for every number n, $H[\overline{n}]$ is provable in $S \leftrightarrow r(h, n) \in \omega_{f(n)} \leftrightarrow H[\overline{n}] \in W_{f(n)}$ (since $r(h, n)$ is the Gödel number of $H[\overline{n}]$).

Corollary. *If S has the sentential recursion property, then there is a formula $H(v_1)$ such that for every n, $H[\overline{n}]$ is provable in S if, and only if, $H[\overline{n}] \in W_n$.*

Proof. By Theorem 1.1, taking the identity function for f.

By Theorems 1 and 1.1 we have

Theorem 1.2. *If S is effectively Gödelian, then there is a formula $H(v_1)$ such that for all n, $H[\overline{n}]$ is provable in S if, and only if, $H[\overline{n}] \in W_n$.*

Remark. Theorem 1.2 above is, of course, a stronger result than Theorem 3 of Chapter 4 (which says only that every effectively Gödelian system is sententially generative). To obtain Theorem 3 of Chapter 4 from Theorem 1.2 above, just take $\sigma(x)$ to be $r(h, x)$.

Exercise 1. Prove that S has the sentential recursion property iff the following condition holds: For every recursive function $f(x)$, there is a formula $H(v_1)$ with Gödel number h which represents $\omega_{f(h)}$ in S.

Exercise 2. Suppose that S has the sentential recursion property. Prove that for any r.e. relation $R(x, y)$ there is a formula $H(v_1)$ with Gödel number h such that for every number n, $H[\overline{n}]$ is provable in S iff $R(n, gnH[\overline{n}])$. [By $gn\,X$ we mean the Gödel number of X.]

Exercise 3. Show that a system S satisfies the conclusion of Ex. 2 iff S satisfies the conclusion of Theorem 1.1.

II.   DSR and Semi-DSR Systems

We now turn to some properties that will prove more significant. We will say that S has the *double sentential recursion property*—in short that S is DSR—if for any disjoint r.e. relations $R_1(x, y)$ and $R_2(x, y)$, there is a number h such that E_h is a formula (in just the free variable v_1) which exactly separates $x : R_1(x, h)$ from $x : R_2(x, h)$.

We shall say that S has the semi-double sentential recursion property—in short, that S is semi-DSR—if for any disjoint r.e. relations $R_1(x, y)$ and $R_2(x, y)$, there is a formula $H(v_1)$ with Gödel number h that strongly (but not necessarily exactly) separates $x : R_1(x, h)$ from $x : R_2(x, h)$. We note that this condition implies (in fact is equivalent to) the condition that for any r.e. relations $R_1(x, y)$

and $R_2(x, y)$ (not necessarily disjoint), there is a formula $E_h(v_1)$ that strongly separates

$$x : R_1(x, h) - x : R_2(x, h)$$

from

$$x : R_2(x, h) - x : R_1(x, h)$$

(because for any r.e. relations $R_1(x, y)$ and $R_2(x, y)$, there exists disjoint r.e. relations $R_1'(x, y)$ and $R_2'(x, y)$ such that $R_1 - R_2 \subseteq R_1'$ and $R_2 - R_1 \subseteq R_2'$).

One reason we are interested in semi-DSR systems is that any such system which is consistent and axiomatizable is uniformly incompletable (as we will see).

§2.   Rosser Systems For Binary Relations.

Theorem 2.

(a) *If S is a Rosser system for binary relations, then S is semi-DSR.*

(b) *If S is an exact Rosser system for binary relations, then S is DSR.*

Suppose S is a Rosser system for binary relations. Let $g(n)$ be the Gödel number of $E_n[v_1, \overline{n}]$ (i.e. of $\forall v_2(v_2 = \overline{n} \supset E_n)$). The function $g(x)$ is recursive and for any number n, $E_{g(n)}$ is the expression $E_n[v_1, \overline{n}]$. If $E_n(v_1, v_2)$ is a formula in the variables v_1 and v_2, then $E_{g(n)}$ is a formula in v_1, and for any number m, $E_{g(n)}[\overline{m}]$ is the sentence $E_n[\overline{m}, \overline{n}]$.

Given any disjoint binary relations $R_1(x, y)$ and $R_2(x, y)$, the relations $R_1(x, g(y))$ and $R_2(x, g(y))$ are r.e. and disjoint. Hence, there is a formula $E_k(v_1, v_2)$ which strongly separates the relation $R_1(x, g(y))$ from $R_2(x, g(y))$; hence $E_k[v_1, \overline{k}]$ strongly separates $x : R_1(x, g(k))$ from $x : R_2(x, g(k))$; hence $E_{g(k)}(v_1)$ (which is $E_k[v_1, \overline{k}]$) effects this separation. Thus $E_h(v_1)$ strongly separates $x : R_1(x, h)$ from $x : R_2(x, h)$, where $h = g(k)$.

Proof of (b). If S is an exact Rosser system for binary relations, then there is some k such that $E_k(v_1, v_2)$ exactly separates $R_1(x, g(y))$ from $R_2(x, g(y))$ (assuming R_1 and R_2 are disjoint) and thus h (viz. $g(k)$) is the Gödel number of a formula that exactly separates $x : R_1(x, h)$ from $x : R_2(x, h)$.

Consistent Axiomatizable Rosser Systems for Binary Relations. It is obvious that if S is DSR, then it is also an exact Rosser system for sets (because if A_1 and A_2 are disjoint r.e. sets, we take $R_1(x,y)$ iff $x \in A_1$, and $R_2(x,y)$ iff $x \in A_2$. Then the relations R_1 and R_2 are r.e. and disjoint and $x : R_1(x,h) = A_1$ and $x : R_2(x,h) = A_2$). Thus, the following result is an extension of Shepherdson's exact separation theorem (Th. S_2, Ch. 0).

Theorem 2'. *If S is a consistent axiomatizable Rosser system for binary relations, then S is DSR.*

Proof. This is an interesting consequence of Theorem S_2^* of Ch. 0. Assume hypothesis. We take the same recursive function $g(x)$ as in the proof of Theorem 2 ($g(n)$ is the Gödel number of $E_n[v_1, \overline{n}]$). Then, given disjoint r.e. relations $R_1(x,y)$ and $R_2(x,y)$, the relations $R_1(x,g(y))$ and $R_2(x,g(y))$ are disjoint and r.e., so by Th. S_2^*, there is a formula $E_k(v_1,v_2)$ such that $E_k[v_1, \overline{k}]$ *exactly* separates $x : R_1(x,g(k))$ from $x : R_2(x,g(k))$. And so $E_{g(k)}(v_1)$ effects this exact separation.

We shall later prove a much stronger result, but for now, Theorem 2' suffices to yield

Theorem R$_2$. *Every consistent axiomatizable extension of (R) has the double sentential recursion property.*

§3. Effective Rosser Systems for Sets.

We recall from Chapter 5 that by a Rosser function for S we mean a recursive function $\pi(x,y)$ such that for all numbers i and j, $\pi(i,j)$ is the Gödel number of a formula $E_{\pi(i,j)}(v_1)$ which strongly separates $\omega_i - \omega_j$ from $\omega_j - \omega_i$ (and, hence, ω_i from ω_j, if these two sets are disjoint). If $E_{\pi(i,j)}$ *exactly* separates (ω_i, ω_j) (provided they are disjoint), then we call $\pi(x,y)$ an *exact* Rosser function for S. We say that S is effectively a Rosser system (effectively an exact Rosser system) for sets, or that S is an effective Rosser system (effective exact Rosser system) for sets, if there is a Rosser function (exact Rosser function) for S.

We proved in Chapter 5 (Th. 8 and Th 8.1) that if S is a Rosser system for binary relations, then it is effectively a Rosser system for sets, and if S is an exact Rosser system for binary relations, then S is effectively an exact Rosser system for sets. And so the following theorem is a strengthening of Theorem 2.

Theorem 3.

(a) *If S is effectively a Rosser system for sets, then S is semi-DSR.*

(b) *If S is effectively an exact Rosser system for sets, then S is DSR.*

Proof. We must now use a weak double recursion theorem in place of a weak recursion theorem.

Suppose $\pi(x, y)$ is a Rosser function for S. Consider any two r.e. relations $R_1(x, y)$ and $R_2(x, y)$. By Theorem 1.1, Chapter 9, there are numbers a and b such that

$$\omega_a = x : R_1(x, \pi(a, b))$$

and

$$\omega_b = x : R_2(x, \pi(a, b)).$$

Now suppose $R_1(x, y)$ and $R_2(x, y)$ are disjoint relations. Then the sets ω_a and ω_b are disjoint.

Proof. Proof of (a) Since $\pi(x, y)$ is a Rosser function for S, then $E_{\pi(a,b)}$ is a formula that strongly separates ω_a from ω_b in S. Hence it strongly separates $x : R_1(x, \pi(a, b))$ from $x : R_2(x, \pi(a, b))$.

Proof. Proof of (b) If $\pi(x, y)$ is an exact Rosser function for S, then $E_{\pi(a,b)}$ exactly separates this same pair.

§4.   Some Stronger Properties.

One of our aims is to show that if S is consistent, axiomatizable and also effectively a Rosser system for sets, then S is uniformly incompletable. For this purpose, Theorem 3(a) is enough. But we have another aim in mind (which will be apparent in the final section of this chapter) for which we will need a strengthening of Theorem 3, to which we now turn.

We will say that S is *effectively* DSR if there is a recursive function $h(x, y)$ (under which S will be said to be effectively DSR) such that for any disjoint r.e. relations $R_i(x, y)$ and $R_j(x, y)$, the number $h(i, j)$ is the Gödel number of a formula that exactly separates $x : R_i(x, h(i, j))$ from $x : R_j(x, h(i, j))$ in S.

We will say that S is effectively semi-DSR under $h(x, y)$ if for any two r.e. relations $R_i(x, y)$ and $R_j(x, y)$ (not necessarily disjoint), $h(i, j)$ is the Gödel number of a formula that strongly separates

$$x : R_i(x, h(i, j)) - x : R_j(x, h(i, j))$$

from

$$x : R_j(x, h(i, j)) - x : R_i(x, h(i, j)).$$

The following theorem is the strengthening of Theorem 3 that we will later need.

Theorem 3*.

(a) *If S is effectively a Rosser system for sets, then S is effectively semi-DSR.*
(b) *If S is effectively an exact Rosser system for sets, then S is effectively DSR.*

Proof. We now need to use a "strong" double recursion theorem (Th. 2.6, Ch. 9).

Proof of (a). Suppose $\pi(x, y)$ is a Rosser function for S. By Th. 2.6, Ch. 9, there are recursive functions $t_1(y_1, y_2)$ and $t_2(y_1, y_2)$ such that for all i and j,

(1) $\omega_{t_1(i,j)} = x : R_i(x, \pi(t_1(i,j), t_2(i,j)))$,
(2) $\omega_{t_2(i,j)} = x : R_j(x, \pi(t_1(i,j), t_2(i,j)))$.

We let $h(x, y) = \pi(t_1(x, y), t_2(x, y))$. Then for any numbers i and j, $h(i, j)$ is the Gödel number of a formula $E_{h(i,j)}$ that strongly separates $\omega_{t_1(i,j)} - \omega_{t_2(i,j)}$ from $\omega_{t_2(i,j)} - \omega_{t_1(i,j)}$ and, hence, strongly separates $x : R_i(x, h(i,j)) - x : R_j(x, h(i,j))$ from $x : R_j(x, h(i,j)) - x : R_i(x, h(i,j))$.

Proof of (b). If $\pi(x, y)$ is an exact Rosser function for S, then if the relations $R_i(x, y)$ and $R_j(x, y)$ are disjoint, so are the sets $\omega_{t_1(i,j)}$ and $\omega_{t_2(i,j)}$. Hence these sets are exactly separated by $E_{h(i,j)}$.

To complete the picture, let us say that S is effectively SR (effectively has the sentential recursion property) if there is a recursive function $h(x)$ such that for every number i, the number $h(i)$ is the Gödel number of a formula which represents $x : R_i(x, h(i))$ in S. Then Theorem 1 has the following strengthening.

Theorem 1*. *If S is effectively Gödelian, then S is effectively SR.*

We also remark that the converses of Theorem 1* and of (a) and (b) of Theorem 3* all hold, and so we have

(1) S is effectively Gödelian if and only if S is effectively SR.
(2) S is effectively a Rosser system for sets if and only if S is effectively semi-DSR.
(3) S is effectively an exact Rosser system for sets if and only if S is effectively DSR.

We leave the proofs of these conditions as exercises that follow.

Exercise 4. Prove Theorem 1* by using Theorem 2.1, Chapter 8.

Exercise 5. Using the iteration theorem, show that there is a recursive function $t(z)$ such that for all i, x and y,

$$R_{t(i)}(x, y) \leftrightarrow x \in \omega_i.$$

Now show that if S is effectively SR under $h(x)$, then S is effectively Gödelian under $h(t(x))$.

Next, show that if S is semi-DSR (DSR) under $h(x, y)$, then $h(tx, ty)$ is respectively a Rosser function (exact Rosser function) for S.

III. Rosser Fixed Point Properties and Uniform Incompletability

We recall that we are letting S_n be that system whose axioms are those of S together with all formulas in W_n. And we are defining S to be uniformly incompletable if there is a formula $H(v_1)$ such that for any n for which S_n is consistent, the sentence $H(\overline{n})$ is an undecidable sentence of S_n. We now turn to an interesting property which is implied by the property of being semi-DSR and which in turn implies uniform incompletability (assuming consistency and axiomatizability).

§5. **Rosser Fixed Point Properties.**   We will say that S has the *Rosser fixed point property* if for any recursive functions $f_1(x)$ and $f_2(x)$, there is a formula $H(v_1)$ such that for every n for which $\omega_{f_1(n)}$ and $\omega_{f_2(n)}$ are disjoint,

(1) $H[\overline{n}] \in W_{f_1(n)} \Rightarrow H[\overline{n}]$ is provable in S.
(2) $H[\overline{n}] \in W_{f_2(n)} \Rightarrow H[\overline{n}]$ is refutable in S.

If, in (1) and (2), we can replace $\Rightarrow$ with $\leftrightarrow$, then we will say that S has the *exact* Rosser fixed point property.

Theorem 4.

(a) *If S is semi-DSR, then S has the Rosser fixed point property.*
(b) *If S is DSR, then S has the exact Rosser fixed point property.*

Proof of (a). Suppose S is semi-DSR. Given recursive functions $f_1(x)$ and $f_2(x)$, we let $R_1(x, y)$, be the relation $E_y[\overline{x}] \in W_{f_1(x)}$ (it is r.e., since it is the relation $r(y, x) \in \omega_{f_1(x)}$) and we let $R_2(x, y)$ be the r.e. relation $E_y[\overline{x}] \in W_{f_2(x)}$. Then there is a formula $H(v_1)$ with Gödel number h that strongly separates

$$x : R_1(x, h) - x : R_2(x, h)$$

from

$$x : R_2(x, h) - x : R_1(x, h).$$

Now suppose $\omega_{f_1(n)}$ and $\omega_{f_2(n)}$ are disjoint. Then

(1) $H[\overline{n}] \in W_{f_1(n)} \Rightarrow H[\overline{n}] \in W_{f_1(n)} - W_{f_2(n)} \Rightarrow$
$R_1(n, h) \wedge \sim R_2(n, h) \Rightarrow H[\overline{n}]$ is provable in S.

(2) Similarly, $H[\overline{n}] \in W_{f_2(n)} \Rightarrow H[\overline{n}]$ is refutable in S.

Proof of (b). Suppose S is DSR. Given recursive functions $f_1(x)$ and $f_2(x)$, we now define $R_1(x, y)$ iff $r(y, x) \in \omega_{f_1(x)}$ before $r(y, x) \in \omega_{f_2(x)}$, and we define $R_2(x, y)$ iff $r(y, x) \in \omega_{f_2(x)}$ before $r(y, x) \in \omega_{f_1(x)}$. The relations $R_1(x, y)$ and $R_2(x, y)$ are disjoint and r.e., so now there is a formula $H(v_1)$ with Gödel number h which exactly separates $x : R_1(x, h)$ from $x : R_2(x, h)$. Now suppose $\omega_{f_1(n)}$ and $\omega_{f_2(n)}$ are disjoint. Then

(1) $H[\overline{n}] \in W_{f_1(n)} \leftrightarrow r(n, h) \in \omega_{f_1(n)} \leftrightarrow r(n, h) \in \omega_{f_1(n)}$ before $r(n, h) \in \omega_{f_2(n)} \leftrightarrow R_1(n, h) \leftrightarrow H[\overline{n}]$ is provable in S.

(2) Similarly, $H[\overline{n}] \in W_{f_2(n)} \leftrightarrow H[\overline{n}]$ is refutable in S.

From Theorems 3 and 4 we have

Theorem 4.1.

(a) *If S is effectively a Rosser system for sets, then S has the Rosser fixed point property.*

(b) *If S is effectively an exact Rosser system for sets, then S has the exact Rosser fixed point property.*

Exercise 6. Show that S is semi-DSR iff the following condition holds: For any recursive functions $f_1(x)$ and $f_2(x)$, there is a formula $H(v_1)$ with Gödel number h which strongly separates $\omega_{f_1(h)} - \omega_{f_2(h)}$ from $\omega_{f_2(h)} - \omega_{f_1(h)}$.

Exercise 7. Show that S has the Rosser fixed point property iff the following condition holds (where by $gnH[\overline{n}]$ we mean the Gödel number of $H[\overline{n}]$): For any r.e. relations $R_1(x, y)$ and $R_2(x, y)$, there is a formula $H(v_1)$ such that for all n,

1. $R_1(n, gn(H[\overline{n}])) \wedge \sim R_2(n, gn(H[\overline{n}])) \Rightarrow H[\overline{n}]$ is provable in $\mathcal{S}$.
2. $R_2(n, gn(H[\overline{n}])) \wedge \sim R_1(n, gn(H[\overline{n}])) \Rightarrow H[\overline{n}]$ is refutable in $\mathcal{S}$.

§6. Uniform Incompletability.   Now we prove

Theorem 5. *If $\mathcal{S}$ is consistent, axiomatizable and has the Rosser fixed point property, then $\mathcal{S}$ is uniformly incompletable.*

Proof. Suppose that $\mathcal{S}$ is axiomatizable. It is then a routine matter to verify that there is a recursive function $f_1(x)$ such that for all n, $W_{f_1(n)}$ is the set of provable formulas of $\mathcal{S}_n$. For any number n, let $neg(n)$ be the Gödel number of the expression $\sim E_n$. The function $neg(x)$ is recursive, and so by the iteration theorem applied to the r.e. relation $neg(x) \in \omega_{f_1(y)}$, there is a recursive function $f_2(x)$ such that for any n, $\omega_{f_2(n)} = x : neg(x) \in \omega_{f_1(n)}$, and so $W_{f_2(n)}$ is the set of refutable formulas of $\mathcal{S}_n$.

Now, suppose $\mathcal{S}$ satisfies the hypothesis. Take any n such that $\mathcal{S}_n$ is consistent. Then the sets $W_{f_2(n)}$ and $W_{f_1(n)}$ are disjoint, and so there is a formula $H(v_1)$ such that $H[\overline{n}] \in W_{f_2(n)}$ implies $H[\overline{n}]$ is provable in $\mathcal{S}$, and $H[\overline{n}] \in W_{f_1(n)}$ implies $H[\overline{n}]$ is refutable in $\mathcal{S}$. And so

1. $H[\overline{n}]$ refutable in $\mathcal{S}_n \Rightarrow H[\overline{n}]$ provable in $\mathcal{S}$,
2. $H[\overline{n}]$ provable in $\mathcal{S}_n \Rightarrow H[\overline{n}]$ refutable in $\mathcal{S}$.

If $H[\overline{n}]$ were provable in $\mathcal{S}_n$, then by (2), it would be refutable in $\mathcal{S}$ and refutable in $\mathcal{S}_n$, and $\mathcal{S}_n$ would be inconsistent. If $H[\overline{n}]$ were refutable in $\mathcal{S}_n$, then by (1) it would be provable in $\mathcal{S}$ and provable in $\mathcal{S}_n$, so again $\mathcal{S}_n$ would be inconsistent. Since $\mathcal{S}_n$ is assumed consistent, then $H[\overline{n}]$ is undecidable in $\mathcal{S}_n$. Hence also $H(\overline{n})$ is undecidable in $\mathcal{S}_n$. And so $\mathcal{S}$ is uniformly incompletable.

By Theorem 4 and Theorem 5 we have

Theorem 6. *If $\mathcal{S}$ is consistent, axiomatizable and semi-DSR, then $\mathcal{S}$ is uniformly incompletable.*

From Theorem 3(a) and Theorem 6 we have one of our principal results.

Theorem 7. *Every consistent axiomatizable effective Rosser system for sets is uniformly incompletable.*

As a corollary we have

Theorem R$_3$ (Pour-El). *Every consistent axiomatizable extension of (R) is uniformly incompletable.*

Discussion. Theorem 7 is a generalization of Th. R$_3$ which has the advantage of being applicable not only to systems couched in the language of first-order logic (as is the system (R)) but to the more abstract representation systems of T.F.S. (which do not involve the logical connectives or quantifiers). In short, Theorem 7, unlike Theorem R$_3$, is not a first-order theorem at all (though it is applicable to first-order systems). Pour-El [1968] gave a very different generalization of Theorem R$_3$, which is essentially of a first-order nature. She considered the smaller class of recursive functions known as *primitive* recursive functions (cf. Kleene, Rogers, Boolos and Jeffrey, or virtually any standard treatment of Gödel's theorem for a definition) and proved the following result (which we will call *Pour-El's Theorem*): If S is any consistent axiomatizable extension of Ω_4 and Ω_5 (cf. §7, Ch. 0) in which all primitive recursive functions of one argument are strongly definable, then S is uniformly incompletable. It is difficult to compare the strength of her hypothesis with that of our Th. 7. We shall shortly prove another generalization of Theorem R$_3$ whose hypothesis is apparently weaker then hers but which appears to be also incomparable in strength with that of Theorem 7.

§7. The Weakened Putnam-Smullyan Conditions.
We are saying that a function $f(x)$ is *admissible* in S if for every formula $H(v_1)$, there is a formula $G(v_1)$ such that for every n, the sentence $G(\overline{n}) \equiv H(\overline{f(n)})$ is provable in S. [This notion of *admissible*, unlike that of a function being strongly definable, does *not* involve the quantifiers]. We proved (Ch. 0, Th. 11.1) that if $f(x)$ is strongly definable in S, then it is admissible in S. As a matter of fact, admissibility is the *only* consequence of strong definability that we ever used in any of our proofs so far. That is to say, in all theorems following Th. 11.1, Ch. 0, if we replaced "strongly definable" by "admissible", the results would still go through. In particular, the Ehrenfeucht-Feferman theorem, the Putnam-Smullyan theorem and its strengthened version in Chapter 7 all go through if we replace "strongly definable" by the weaker, but more generally applicable "admissible". [Indeed, in each of our proofs, we had to appeal to Th. 11.1, Ch. 0 each time!]

And now we shall say that S satisfies the *weakened* Putnam-Smullyan conditions if S is a consistent axiomatizable Rosser system for sets such that for every number h, the function $J(h,x)$ is admissible in S.

We will prove

Theorem 8. *If S satisfies the weakened Putnam-Smullyan conditions, then S is uniformly incompletable.*

To prove Theorem 8, we first need a strengthening of Theorem 5. Let us say that S *almost* has the Rosser fixed point property if for any recursive functions $f_1(x)$ and $f_2(x)$, there is a formula $H(v_1)$ and a function $\varphi(x)$ which is admissible in S and such that for any n for which $\omega_{f_1(n)}$ and $\omega_{f_2(n)}$ are disjoint, the following conditions hold:

1. $H[\overline{\varphi(n)}] \in W_{f_1(n)} \Rightarrow H[\overline{\varphi(n)}]$ is provable in S,
2. $H[\overline{\varphi(n)}] \in W_{f_2(n)} \Rightarrow H[\overline{\varphi(n)}]$ is refutable in S.

[Of course this condition is a weakening of the Rosser fixed point property, since the identity function is obviously admissible in S.]

Now suppose S is consistent, axiomatizable and almost has the Rosser fixed point property. Then it must be uniformly incompletable by the following argument: We take $f_1(x)$ and $f_2(x)$ as in the proof of Theorem 5, and we then take $H(v_1)$ and an admissible function $\varphi(x)$ satisfying (1) and (2) above (for the particular functions $f_1(x)$ and $f_2(x)$). Then by an argument similar to part of the proof of Theorem 5, we see that for any n for which S_n is consistent, the sentence $H(\overline{\varphi(n)})$ is undecidable in S_n. But since $\varphi(x)$ is admissible in S, there is a formula $G(v_1)$ such that for all n, the sentence $G(\overline{n}) \equiv H(\overline{\varphi(n)})$ is provable in S (and, hence, in every extension of S), and so $G(\overline{n})$ is undecidable in S_n (if S_n is consistent.) And so we have

Theorem 5*. *If S is consistent, axiomatizable and almost has the Rosser fixed point property, then S is uniformly incompletable.*

Now we can prove Theorem 8. Assume hypothesis. We will show that S almost has the Rosser fixed point property and then by Th. 5*, S is uniformly incompletable.

Given recursive functions $f_1(x)$ and $f_2(x)$, we let A_1 be the set of all numbers $J(x,y)$ such that $E_x[\overline{J(x,y)}] \in W_{f_1(y)}$ and we let A_2 be the set of all numbers $J(x,y)$ such that $E_x[\overline{J(x,y)}] \in W_{f_2(y)}$. Then there is a formula $E_h(v_1)$ that strongly separates $A_1 - A_2$ from

$A_2 - A_1$. Now, let n be any number for which $\omega_{f_1(n)}$ and $\omega_{f_2(n)}$ are disjoint. Then,

1. $E_h[\overline{J(h,n)}] \in W_{f_1(n)} \Rightarrow E_h[\overline{J(h,n)}] \in W_{f_1(n)} - W_{f_2(n)} \Rightarrow$ $(h,n) \in A_1 - A_2 \Rightarrow E_h[\overline{J(h,n)}]$ is provable in S.
2. Similarly, $E_h[\overline{J(h,n)}] \in W_{f_2(n)}$ implies that $E_h[\overline{J(h,n)}]$ is refutable in S.

And so we take $\varphi(x) = J(h,x)$. By hypothesis, $\varphi(x)$ is admissible in S. This concludes the proof.

Discussion. The hypothesis of Theorem 8 appears to be of incomparable strength with that of Theorem 7, but it also seems to be weaker than Pour-El's hypothesis for the following reasons: It is a well-known result of recursive function theory that every r.e. set is the range of a *primitive* recursive function and, hence, if all primitive recursive functions are definable in S, then all r.e. sets are *enumerable* in S (in the sense of §2, Ch. 0). Hence if S is an extension of Ω_4 and Ω_5, then S is a Rosser system for sets (cf. Separation Lemma, §8, Ch. 0). Also, for any number h, the function $J(h,x)$ is a primitive recursive function, and so any system satisfying Pour-El's hypothesis also satisfies the hypothesis of Theorem 8.

From Theorem 8, of course, follows the weaker result.

Theorem 8°. *If S is a consistent axiomatizable Rosser system for sets in which all recursive functions of one argument are admissible, then S is uniformly incompletable.*

In some of the exercises that follow, we outline another proof of the weaker Theorem 8°, because it brings to light some facts about doubly generative pairs that are of interest on their own account.

Exercise 8. If in our definition of a system almost having the Rosser fixed point property, we can replace "$\Rightarrow$" by "$\leftrightarrow$", then we will say that S almost has the *exact* Rosser fixed point property. Using Theorem $S_2^\sharp$ of Chapter 7, prove that if S satisfies the hypothesis of Theorem 8, then S almost has the exact Rosser fixed point property.

Exercise 9. Now for some facts about doubly generative pairs. Suppose (A_1, A_2) is D.G. Prove that for any recursive functions $h(x)$, $f_1(x)$ and $f_2(x)$, there is a recursive function $g(x)$ such that for all n for which $\omega_{f_1(n)}$ and $\omega_{f_2(n)}$ are disjoint,

1. $g(n) \in A_1 \leftrightarrow h(g(n)) \in \omega_{f_1(n)}$;
2. $g(n) \in A_2 \leftrightarrow h(g(n)) \in \omega_{f_2(n)}$.

[Hint: Let $\phi(x, y)$ be a D.G. function for (A_1, A_2). Then take a recursive function $t(y)$ such that for all $i : \omega_{t(i)} = h^{-1}(\omega_i)$. Then take $g(x) = \phi(tf_1x, tf_2x)$ and show that $g(x)$ works.]

Exercise 10. Suppose $h(x)$ is a recursive function that reduces a D.G. pair (A_1, A_2) to a pair (α_1, α_2). Using Exercise 9, show that for any recursive functions $f_1(x)$ and $f_2(x)$, there is a recursive function $g(x)$ such that for any n for which $\omega_{f_1(n)}$ and $\omega_{f_2(n)}$ are disjoint,

(1) $h(g(n)) \in \alpha_1 \leftrightarrow h(g(n)) \in \omega_{f_1(n)}$;
(2) $h(g(n)) \in \alpha_2 \leftrightarrow h(g(n)) \in \omega_{f_2(n)}$.

Exercise 11. Now show that if some D.G. pair (A_1, A_2) is strongly separable in S and if all recursive functions are admissible in S, then S almost has the Rosser fixed point property (in fact S has the stronger property that there is a single formula $H(v_1)$ such that for any recursive functions $f_1(x)$ and $f_2(x)$, there is an admissible function $\varphi(x)$ such that for all n for which $\omega_{f_1(n)}$ and $\omega_{f_2(n)}$ are disjoint, $H[\overline{\varphi(n)}]$ is in $W_{f_1(n)}(W_{f_2(n)})$ implies $H[\overline{\varphi(n)}]$ is provable (respectively refutable) in S). [Theorem 8° then follows from this and Th. 5*.]

IV. Finale

Now we shall prove the following two results (which are apparently of incomparable strength).

Theorem 9. *If S is consistent, axiomatizable and semi-DSR, then S is an exact Rosser system for sets.*

Theorem 10. *If S is consistent, axiomatizable and effectively semi-DSR, then S is effectively an exact Rosser system for sets.*

When we have proved Theorem 9, then by Theorem 3(a) we will have

Theorem I. *If S is consistent, axiomatizable and effectively a Rosser system for sets, then S is an exact Rosser system for sets.*

This is a strengthening of Shepherdson's exact separation theorem, since any Rosser system for binary relations is effectively a Rosser system for sets. Also, our proof of Theorem I yields a new proof of Shepherdson's theorem which combines a variant of Shepherdson's argument with the use of the weak double recurson theorem (needed for the proof of Theorem 3(a)).

When we have proved Theorem 10, then by Theorem 3*(a) we will have the still stronger result.

Theorem I*. *If S is consistent, axiomatizable and also effectively a Rosser system for sets, then S is effectively an exact Rosser system for sets.*

Remarks. Just because a system is both an effective Rosser system for sets and an exact Rosser system for sets, doesn't mean that it must be effectively an exact Rosser system for sets. Thus Theorem I* is indeed a substantial strengthening of Theorem I.

Now for the proofs.

Proof of Theorem 9:   Suppose S is consistent, axiomatizable and semi-DSR. Given disjoint sets A and B, we let $R_1(x,y)$ and $R_2(x,y)$ be the following r.e. relations:

$$R_1(x,y) \text{ iff } x \in A \vee (E_y[\overline{x}] \text{ is refutable in } S).$$

$$R_2(x,y) \text{ iff } x \in B \vee (E_y[\overline{x}] \text{ is provable in } S).$$

Then by hypotheses there is a formula $E_h(v_1)$ that strongly separates $x : (R_1(x,h) \wedge \sim R_2(x,h))$ from $x : (R_2(x,h) \wedge \sim R_1(x,h))$. Then for any n we have

1. $[(n \in A \vee E_h[\overline{n}] \text{ is refutable}) \wedge \sim (n \in B \vee E_h[\overline{n}] \text{ is provable})]$
   $\Rightarrow E_h[\overline{n}]$ is provable.
2. $[(n \in B \vee E_h[\overline{n}] \text{ is provable}) \wedge \sim (n \in A \vee E_h[\overline{n}] \text{ is refutable})$
   $\Rightarrow E_h[\overline{n}]$ is refutable.

Since A and B are disjoint and, by the assumption of consistency, $E_h[\overline{n}]$ is not both provable and refutable, it follows from (1) and (2) by propositional logic that $E_h[\overline{n}]$ is provable iff $n \in A$, and refutable iff $n \in B$.

Remarks. It is of interest to compare the relations $R_1(x,y)$ and $R_2(x,y)$ that we used with those of Shepherdson ($x \in A \vee E_y[\overline{x},\overline{y}]$ is refutable; $x \in B \vee E_y[\overline{x},\overline{y}]$ is provable). His relations built a diagonalization right within them; ours did not, since the weak double recursion theorem contained within it all the diagonalization that was needed.

Proof of Theorem 10:   Suppose that S is consistent and axiomatizable and that S is effectively semi-DSR under $h(x,y)$. The relation

"$x \in \omega_z \vee E_y[\overline{x}]$" is refutable, and the relation "$x \in \omega_z \vee E_y[\overline{x}]$" is provable are both r.e. (since $\mathcal{S}$ is axiomatizable). So by the iteration theorem, there are recursive functions $t_1(z)$ and $t_2(z)$ such that for all i

1. $R_{t_1(i)}(x, y) \leftrightarrow (x \in \omega_i \vee E_y[\overline{x}]$ is refutable),
2. $R_{t_2(i)}(x, y) \leftrightarrow (x \in \omega_i \vee E_y[\overline{x}]$ is provable).

We take $\pi(x, y) = h(t_1(x), t_2(y))$. Then for any i and j, $E_{\pi(i,j)}(v_1)$ strongly separates $x : (R_{t_1(i)}(x, \pi(i,j)) \wedge \sim R_{t_2(j)}(x, \pi(i,j)))$ from $x : (R_{t_2(j)}(x, \pi(i,j)) \wedge \sim R_{t_1(i)}(x, \pi(i,j)))$ in $\mathcal{S}$. Thus conditions (1) and (2) of the proof of Theorem 9 hold when we replace "h" by "$\pi(i,j)$", "A" by "ω_i" and "B" by "ω_j". And so if ω_i and ω_j are disjoint, then $E_{\pi(i,j)}(v_1)$ exactly separates ω_i from ω_j. Thus $\pi(x, y)$ is an exact Rosser function for $\mathcal{S}$.

Having proved Theorem 10, then by Theorem 3*(a), our proof of Theorem I* is complete.

We note that Theorem I* and Theorem 3*(b) yield

Corollary. *If a consistent axiomatizable system is effectively semi-DSR, then it is effectively DSR.*

And so we see that for a consistent axiomatizable system $\mathcal{S}$, the following four conditions are all equivalent:

(1) $\mathcal{S}$ is effectively a Rosser system for sets.
(2) $\mathcal{S}$ is effectively semi-DSR.
(3) $\mathcal{S}$ is effectively an exact Rosser system for sets.
(4) $\mathcal{S}$ is effectively DSR.

In conclusion, let us summarize some results of Chapter 6 and this Chapter with

Theorem ER. *Every consistent axiomatizable effective Rosser systems for sets enjoys the following properties:*

(1) *The pair (P, R) of its nuclei is completely E.I., doubly generative and doubly universal.*
(2) $\mathcal{S}$ *is uniformly incompletable.*
(3) $\mathcal{S}$ *is effectively an exact Rosser system for sets.*
(4) $\mathcal{S}$ *effectively has the double sentential recursion property.*
(5) $\mathcal{S}$ *has the exact Rosser fixed point property.*

This seems to be an appropriate resting place. Recursion theorems and Shepherdson-type constructions have been two of our major tools. Many of the results we have proved about first-order systems and recursion theory can be unified using the representation systems of T.F.S.—Moreover, fixed point theorems in these areas can be further unified with fixed point theorems in combinatory logic. This is a topic unto itself and will be pursued in our companion volume *Diagonalization and Self-Reference.*

References

[1] Paul Bernays. Review of myhill [1955]. *J. Symbolic Logic* 22:73–76, 1957.

[2] A. Markov. The theory of algorithms. *Tr. Mat. Inst.*, XLII:375, 1961.

[3] Emil Post. Recursively enumerable sets of positive integers and their decision problems. *Bull. Am. Math. Soc.*, 50:284–316, 1944.

[4] Marian Boykan Pour-El. Effectively extensible theories. *J. Symbolic Logic* 33:56–68, 1968.

[5] Hilary Putnam. Decidability and essential undecidability. *J. Symbolic Logic* 22:39–54, 1957.

[6] H. Gordan Rice. Classes of recursively enumerable sets and their decision problems. *Transactions of the American Mathematical Society* (74):358–366, 1953.

[7] Hartley Rogers. **Theory of Recursive Functions and Effective Computability**. McGraw-Hill, 1967.

[8] J.R. Shoenfield. Undecidable and creative theories. *Fundamenta Mathematicae* (XLIX):171–179, 1961.

[9] John Shepherdson. Representability of recursively enumerable sets in formal theories. *Archiv für Mathematische Logik und Grundlagenforshung* 119–127, 1961.

[10] Raymond M. Smullyan. Undecidability and recursive inseparability. *Zeitschr. f. math. Logik und Grundlagen d. Math.* 4:143–147, 1958.

[11] Raymond M. Smullyan. Theories with effectively inseparable nuclei. *Zeitschr. f. math. Logik und Grundlagen d. Math.* 6:219–224, 1960.

[12] Raymond M. Smullyan. Creativity and effective inseparability. *Transactions of the American Mathematical Society* 109(1):135–145, 1963.

[13] Raymond M. Smullyan. Pseudo-uniform reducibility. *Journal of the Mathematical Society of Japan* 15(2):129–133, 1963.

[14] Raymond M. Smullyan. **To Mock a Mocking Bird**. Alfred A. Knopf, 1985.

[15] Raymond M. Smullyan. **Gödel's Incompleteness Theorems**. Oxford University Press, 1992.

[16] A. Tarski. **Undecidable Theories**. North Holland, 1953.

[17] Alfred Tarski. Der Wahrheitsbegriff in den formalalisierten Sprachen. *Studio Philos* 1:261–405, 1936.

[18] Alfred Tarski. A simplified formalization of predicate logic with identity. *Arch. f. Math. Logik und Grundl.* 7:81–101, 1965.

[19] Alan M. Turing. On computable numbers with an application to the ensheidungs problem. *Proc. of the London Math. Soc.* 42(2):230–265, 1936.

[20] Alan M. Turing. On Computable numbers with an application to the ensheidungs problem. *Proc. of the London Math. Soc.* 3(2):544–546, 1936.

Index